Nanotechnology in the Defense Industry

Scrivener Publishing
100 Cummings Center, Suite 541J
Beverly, MA 01915-6106

Advances in Nanotechnology & Applications

Series Editor: Madhuri Sharon

The unique properties of nanomaterials encourage the belief that they can be applied in a wide range of fields, from medical applications to electronics, environmental sciences, information and communication, heavy industries like aerospace, refineries, automobile, consumer and sports good, etc.

This book series will focus on the properties and related applications of nanomaterials so as to have a clear fundamental picture as to why nanoparticles are being tried instead of traditional methods. Since nanotechnology is encompassing various fields of science, each book will focus on one topic and will detail the basics to advanced science for the benefit of all levels of researchers.

Series Editor: Madhuri Sharon, Director, Walchand Centre for Research in Nanotechnology & Bionanotechnology
W.H. Marg, Ashok Chowk, Solapur 413 006
Maharashtra,
India

E-mail:sharonmadhuri@gmail.com

Publishers at Scrivener
Martin Scrivener (martin@scrivenerpublishing.com)
Phillip Carmical (pcarmical@scrivenerpublishing.com)

Nanotechnology in the Defense Industry

Advances, Innovation, and Practical Applications

by

Madhuri Sharon, Angelica Silvestre Lopez Rodriguez, Chetna Sharon and Pio Sifuentes Gallardo

Scrivener Publishing

WILEY

This edition first published 2019 by John Wiley & Sons, Inc., 111 River Street, Hoboken, NJ 07030, USA and Scrivener Publishing LLC, 100 Cummings Center, Suite 541J, Beverly, MA 01915, USA

For more information about Scrivener publications please visit www.scrivenerpublishing.com.

Wiley Global Headquarters
111 River Street, Hoboken, NJ 07030, USA

For details of our global editorial offices, customer services, and more information about Wiley products visit us at www.wiley.com.

Library of Congress Cataloging-in-Publication Data

ISBN 978-1-119-46012-1

Cover image: Pixabay.Com
Cover design by Russell Richardson

Set in size of 13pt and Minion Pro by Manila Typesetting Company, Makati, Philippines

10 9 8 7 6 5 4 3 2 1

Contents

Preface

More than half the globe is confronted with war or war-like volatility and complex security situations. Scientists, including nanotechnologists, in every sphere are concerned and want to contribute. In this book, the challenges of our time, including new demands on the military at every level, are addressed. The need to provide defense personnel with practical tools necessary to succeed in this unpredictable and unstable world is of utmost importance, because the character of warfare is changing with time. This book articulates various areas of the defense support system that can be improved with the incorporation of nanotechnology, as well as ways to further develop a modern, future-oriented, attractive and competitive military support system.

With these concerns in mind, the chapters of this book are devoted to various areas of nanotechnological innovation afforded soldiers such as smart clothing and combat suits, medical therapeutic and diagnostic support, improved air and ground vehicles, portable energy sources, and nanosensors. Nanotechnology supported stealth and counter-stealth technologies, radar absorbing materials, the role of computers in aid of defense and textiles to help in camouflaging are also discussed. In addition, chapters have also been devoted to chemical and biological warfare.

Madhuri Sharon
Mumbai
August 2019

Foreword

The old adage, "With age comes wisdom and experience," holds true for the main author of this book, whom I have known for six years. My first encounter with her was in a spiritual meeting where I noticed her exceptional competency in executing holistic research, as she can inclusively think beyond the benefit of any particular organization or person attributed to her belief in spirituality. Dr. Madhuri Sharon epitomizes the life-long learner, always seeking novel approaches and skills. I have always loved and appreciated her dedication and significant contributions to the revolutionary field of nanotechnology, which is gaining considerable momentum in areas ranging from biomedical applications to their use in modern military systems.

The popular statement, "Small is the new big," is reflected in the enormous amount of applications of nanotechnology in the military domain. In this book, the authors present the advanced features of nanotechnology that can be well integrated into current military tactics used to fight wars. Nanotechnology provides multifunctional materials such as nanobattlesuits, nanoweapons, nanosensors, stealth and camouflage devices that facilitate effective protection and better battlespace awareness. Miniaturization of already existing military equipment to make it smaller, lighter weight and easily concealable can provide significant strategic advantages over the enemy.

This book fills the innovation gap that puts soldiers in peril and highlights the fact that an understanding of nanotechnology is necessary for technological advancements in the defense industry to occur in order to ensure the future security of countries.

Dr. Sushil Chandra
Scientist 'G' and Head of the Biomedical
Engineering Department,
Institute of Nuclear Medicine & Allied Sciences (INMAS)
and Defence Research and Development Organisation
(DRDO), New Delhi, India
July 2019

1

Nanotechnology's Entry into the Defense Arena

Madhuri Sharon

Walchand Centre for Research in Nanotechnology & Bionanotechnology, WCAS, W.H. Marg, Ashok-Chowk, Solapur, Maharashtra, India

Science is based solely on doubt-based, disinterested examination of the natural and physical world. It is entirely independent of personal belief. There is a very important, fundamental concomitant - that is to accept absolutely nothing whatsoever, for which there is no evidence, as having any fundamental validity.
Nobel Laureate Sir Harold W. Kroto

1.1 Introduction

When nanotechnology emerged as a new science, it had attracted the attention of scientists from all branches of science the world over. Similarly, people from different domains of life and governance started realizing the impact of nanotechnology. Authorities from the military, security and defense arena also started thinking about the use of this technology. It was realized that nanotechnology would make a big difference in areas such as troop functionality, high-speed conveyance and

Madhuri Sharon, et al. Nanotechnology in the Defense Industry: Advances, Innovation, and Practical Applications, (1–36) © 2019 Scrivener Publishing LLC

capacity, safety of soldiers, betterment of aircraft, systems for command, control, communication and surveillance, automation and robotics, innovative sensors, advanced war fighter and battle systems capability, electrochemical power, such as batteries or fuel cells, and many more. Nanotechnology is seen as being a technology of national importance to the economy and security by many developed countries especially in the defense sectors. This chapter is about the area of defense that is being impacted by nanotechnology. Almost all countries across the globe are experimenting to understand how this Small Science will help protect their country.

1.2 What Is Nanotechnology

Let us very briefly look at this topic. One may write volumes on what nanotechnology is. But that would be beyond the scope of this book, because this book is about the specific application of nanotechnology in defense. However, in a nut shell, the emergence of nanotechnology, which is a science that deals with the properties of materials with dimensions of 1–100 nm (Figure 1.1), is a result of the fruitful thinking of the visionary

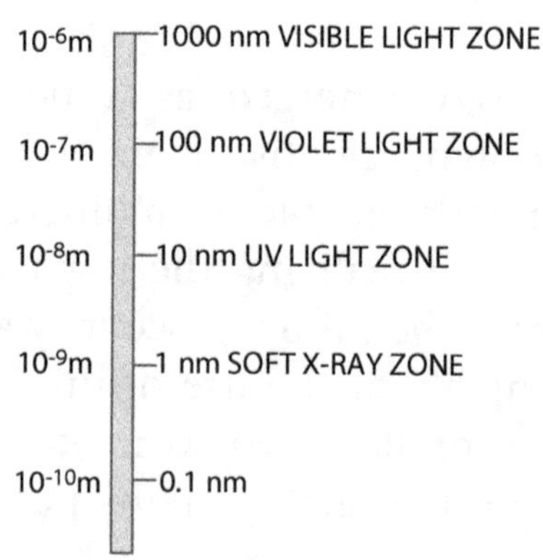

Figure 1.1 Nanometer scale.

Nobel laureate Professor Richard Feynman. Nanotechnology has revolutionized almost every field such as medicine, electronics, information technology, geology, space technology, material science, etc. Enough evidence on this subject has been provided to be able to state that *at macroscale and microscale objects behave differently than that at nanoscale.* The behavioral as well as the structural properties of a nano-object change. When one starts reducing the size of a solid from micro-level to nano-level, its appearance completely changes, especially where metals are concerned. Gold appears lustrous yellow at macroscale, but when brought to nano-level its color turns to red or blue depending on the size.

Such alterations in properties of nanomaterials have created vast applications in Physics, Chemistry, Materials Science, Computers, Engineering, Medical and Biosciences, etc. No wonder venturing into the tiny domain of atoms and molecules is catching on. The main aim and objectives of Nanotechnology and Nanoscience is to unravel as well as comprehend the appearance of strange but useful traits in nanoparticles when they venture into the realm of nanoscale (1–100 nm). The emergence of the physicochemical and optoelectronic properties of nanoparticles is primarily due to confinement of electrons within particles of dimensions smaller than the bulk electron delocalization length. This ability to tune the optical absorption or emission properties (quantum dots) by simple variation in nanoparticle size is particularly attractive in the facile band-gap engineering of material and the growth of quantum dots.

In a nutshell, nanotechnology is the branch of technology that deals with dimensions and tolerances of less than 100 nanometers, especially the manipulation of individual atoms and molecules, or is the manipulation of matter on an atomic, molecular, and supramolecular scale. Such small engineered materials exhibit unique properties that are not found in their bulk counterparts.

The unique properties of nanomaterials that are being utilized for various applications are their chemical reactivity, mechanical strength, electrical properties, optical properties, thermal properties, and durability and quantum effects.

It is difficult to say which country's defense department was first to start working out the possibility of using nanotechnology. However, strong participation of the United States Department of Defense in areas such as sensitive spectrum analyzer 3Hz – 50 GHz, quantum dots for nanoelectronics and sensors, highest Q mechanical resonator (cantilever, nanowire) for chemical sensing, nanomagnetic materials for DNA detection, nanoscale aluminium particles for energetic materials, smallest individual organic light emitting diodes (OLED) device, long shelf life packaging for food (nanoclay PE composite), high strength fibers from twisted nanotube yarns and maskless lithographic process for semiconductors are evidence that they have been investing in working out the applications of nanotechnology.

Out of the Asian countries, Japan is the leading up-and-coming nation investing in nanotechnology, followed by China and Taiwan.

In European countries, Germany, Russia, the United Kingdom, the Netherlands, etc., are running programs in nanotechnology.

1.3 Nanotechnology Offers Innovative Opportunities for Defense

Initially, many thought that nanotechnology applications offered more hype than hope. But as many fantasies started becoming realities, the considerations and joint as well as separate efforts of scientists have reached a consensus that nanotechnology can provide strength and enhanced efficacy to many requirements of the defense arena. The area in which nanotechnology inputs are being envisaged as a must are its applications for (i) soldiers, (ii) security, (iii) weapons, (iv) aeronautics, (v) naval vessels, (vi) vehicles, (vii) satellites and (viii) logistics.

1.4 Nanotechnology for Soldiers

At the moment the maximum focus for the safety of soldiers is on developing a nano combat suit, which should be as thin as spandex and contain health monitors and communications equipment, acts as a shield from bullets, and can protect from biological and chemical attacks. This is for the well-being and security of soldiers. Nanotechnology is crucial for such development because without miniaturization such functionalities cannot be adapted to lightweight, wearable systems.

1.4.1 Smart Clothing Using Nanotechnology for Various Applications

Using the process of manipulating materials on an atomic or molecular scale, clothes are manufactured that have the following characteristics:

- Because it has *greater tolerance to temperature* change, nanotechnology-based fabric can warm or cool the body.
- It is possible to *charge a phone* using a fabric that is solar-powered and to recharge a cell phone or iPod. The change in temperature can open or tighten the weaving pattern of the fabric because the fibers of the fabrics can be modified so their physical properties change with temperature.
- *Anti-microbial nanofabrics* coated with antimicrobial nanoparticles can be used to protect soldier's bodies from microbial infection.

- Another possibility that is being envisaged is to use fabrics coated with *encapsulated insecticides to eliminate mosquitoes.*
- In case a soldier faces smog or toxic gas, scientists are working on fabrics loaded with encapsulated gases that can be released to combat the condition.
- *High-strength fiber* made up of a composite of carbon nanotube impregnated in polymer can be used in nonwoven mats.
- *Multi-Functional Textiles* is another fantasy for the protection of soldiers that is being considered by modifying and functionalizing the surface properties by mixing yarn with nanomaterial and then weaving fabric from it. The new properties that can be exploited will be for the purposes of *sensing,* energy harvesting and energy storage.

1.4.2 Invisibility and Adaptive Camouflage

One of the ongoing dreams of scientists that has been dramatized and presented in many works of fiction and movies is a disappearing or invisible man. For this, one approach that is being considered is to make a cloak or fabric by *manipulating light* so that the soldiers wearing it cannot be seen or disappear. An "ultrathin invisibility skin cloak for visible light" is an outcome of this idea [1]. It is also known as *metamaterial cloaking* as it uses metamaterials, along with manipulating the paths traversed by light through a novel optical material. Metamaterials direct and control the propagation and transmission of specified parts of the light spectrum and render the object invisible.

This result is based on the fact that how we see any object depends on the role of light that bounces off the object and becomes distorted. This distortion is due to the angles and curves of the object. However, if the object is smaller than the wavelength of visible light (300–700 nm), it passes straight through and cannot be seen. That is why even after adding a very high magnification a nanoparticle cannot be seen under the light microscope. The development of a cloak that makes the object invisible is based on this principle. To make a cloak material look invisible it should be fully covered with tiny gold blocks of different nanosizes. Though it seems beyond reach today, soon it will become a reality; especially with the recent report of a dielectric "surface wave cloak" from Hao's group [2], this fantasy seems to be a possibility in the near future. They have made a cloak from engineered gradient index materials (seven ultrathin layers that each had different electric properties) using nanocomposites to control surface wave propagation through advanced additive manufacturing and coated a curved metal plate surface, about the size of a tennis ball, with their cloak. The seven ultrathin layers can hide whatever object they cloak from interacting with electromagnetic waves. Otherwise, the electromagnetic waves would bump into the object and scatter in different directions.

This device is designed analytically and validated through numerical simulations and measurements, showing good agreement and performance as an effective surface wave cloak. The unique aspect of this "surface wave cloak" is that it can make curved surfaces appear flat when they come in contact with electromagnetic waves. The cloak was made out of a gradient-index material with seven ultrathin layers that each had different electric properties. This technology can also be applied to escape detection by radiated probing.

Adaptive camouflage is another approach which is focusing on *electrochromic camouflage*, i.e., using nanoparticle coated fabric which changes colors instantly to blend in with the surroundings. The change in color of the fabric is based on reflective electrochromic device. Adaptive camouflage changes color in response to the environment. This chameleon sheet can be used for draping tanks, vehicles or even as soldier's suit.

In their patented work, Gorodetsky and Xu explain that this device consists of primary color tone plaque that is the main color of the entire fabric and secondary color tone plaque which is patched on this fabric. The fabric is a flexible conductive textile with ion storage layer, an electrolyte layer, the main color camouflage patched of conductive fabric, electrochromic polymer layer and a transparent protective film layer and a conductive flexible conductive fabric camouflage fabric respectively. It is connected to a positive and negative power supply through a wire. Since it uses two fabric electrodes, it makes fabric wearable. Moreover, the electrochromic polymer and a reflective structure having low voltage can be artificially controlled between two or more camouflage patterns that change color fast.

Luminex using integrated glass fibers + LED source is making luminescent fibers by weaving fiber optics into fabric. Super bright LEDs are attached to these colorless fibers that are controlled by a microchip (powered with a battery), which instructs them to emit different colors. An optical sensor is added; hence, the clothes glow by constantly changing colors to match the surroundings.

1.4.3 Armor Fabric

Especially for defense personnel, soldier's antiballistic and shatterproof armor will be of great benefit. It will not only benefit the military, but also fire fighters, police officers, and other emergency responders. Present-day armor, i.e., *Smart Armour,* is not only for protection but is also being researched using nanotechnology so that it can sense the impact of a bullet or shrapnel and automatically stiffen, which should make it even more resistant to penetration and less cumbersome than the ceramic-plate armor troops now wear. Researchers are working on the following materials for use as armor.

1.4.3.1 *Artificial Muscles*

A concept of weaving artificial muscles into armor that could enable soldiers to leap tall walls is being considered. This material uses nanotechnology and electricity and flexes when jolted by electricity and then relaxes when the electricity is turned off.

1.4.3.2 *Strong, Lightweight and Self-Repairing Material*

Nature has always been the guiding force for scientist for many discoveries. The use of abalone sea snails is one such guiding idea for scientists in their search for strong and lightweight armor. The shells of abalone are strong, lightweight and can self-repair. For this purpose, improved body armor using SiO or TiO_2 nanoparticles embedded in epoxy matrix is being explored. SiO nanoparticles in a liquid polymer harden on ballistic impact (*shear thickening fluid*). Similarly, iron nanoparticles dispersed in inert oil harden on stimulation with an electrical pulse (*magnetorheological fluid*).

1.4.3.3 *Tungsten as Ultrastrong Material*

An attempt is currently being made to use tungsten as a basic material since it is five times stronger than steel and twice as strong as currently used impact-resistant material. It is to be used as a new nanomaterial for body armor so as to give ballistic protection. It will be useful for vehicle armor, shields, helmets, and protective enclosures.

1.4.3.4 *Carbon Nanomaterials*

Carbon nanomaterials (CNMs), especially those having onion-like layering or are multiwalled, i.e., multiwall carbon nanotubes (MWCNTs) (Figure 1.2), have also shown strong and durable properties when embedded in polymers. Investigations are underway to make nanosized umbrellas that open to seal the cloth's

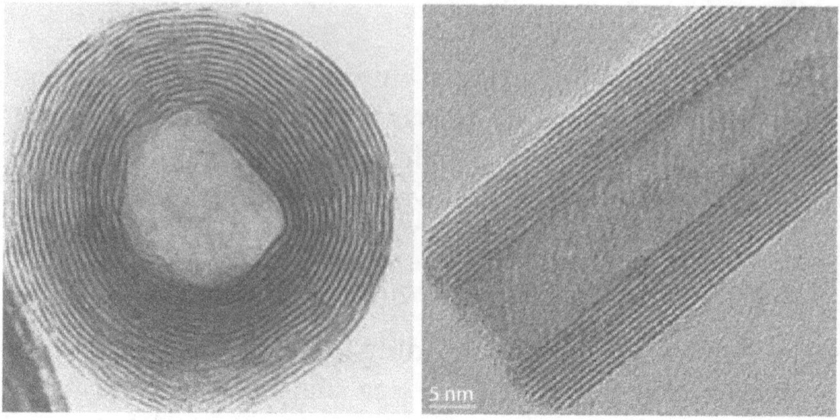

Figure 1.2 TEM images of a multiwall carbon nanotube from two different angles showing lumen in the center surrounded by many walls.

pores, making it impervious to airborne chemicals and pathogens. That would be much easier and lighter than the current equipment required for protecting against biochemical warfare.

1.4.3.5 Future Combat Suits

Future combat suits incorporating the above-mentioned developments in nanotechnology- and nanoelectronics-based textiles will protect against various threats (bullets, grenade fragments, bioagents, chemical agents), will be light, and will monitor vital signs with printed and embedded physiological sensors on textiles (i.e., ECG (electrocardiography) sensors to monitor heart rate, respiration, temperature, dehydration). These suits will have a power grid and energy generating system using ZnO nanowires on textile fibers which will act as piezoelectric nanogenerators. To face extreme temperature conditions, thermal management technologies for textiles use carbon nanofibers and CNTs for heat conductance, ventilation, insulation and local cooling. They will have external sensors for communication and identification. Adaptive camouflage has already been discussed above.

Giving smart armor properties to strong, lightweight backpacks and shoes is also being considered.

Protein nanospheres have been developed at Tel Aviv University. This material is transparent and can self-assemble. It has better properties than both stainless steel and Kevlar. This material could be used in body armor, medical implants, and to strengthen existing composites and ceramics.

1.4.4 Faster Intensive Medical Help

Medical help for a soldier needs a multidirectional approach. Improved medical and casualty care for soldiers is one of the main aims of military research into nanotechnology to produce lightweight, strong and multifunctional materials for use in clothing, both for protection and to provide enhanced connectivity.

1.4.4.1 Diagnostic Support Using Nanotechnology

For monitoring vital signs, such as heart rate or brain signals, of soldiers, sensors and antennas for communicating information from the sensors can be woven into the fabric of shirts or even pillows to use for a wounded soldier. These sensors would also provide the soldier's location to medics via radio. An adhesive chest patch fitted with sensors and a tiny radio can also be used if they are not embedded in the textiles. Another futuristic approach will be to develop medical nanobots and nano-enhanced reconnaissance and communication devices, such as micro-radar, for faster intensive medical help.

1.4.4.2 Nano-Tourniquet

To help a soldier with an injured arm or leg, scientists have developed nanofibers (that are woven into the fabric of the uniform). This fabric will constrict into a tourniquet and thus save blood loss till wounded soldiers can be treated.

1.4.4.3 Antitoxin Guard

Antitoxin guard is made of thin packaging films using nanotechnology that is capable of detecting pathogenic microorganisms, which can cause illness or disease.

1.4.4.4 Lab-on-Chip

For faster medical help and support, the concept of *lab-on-a-chip* is being worked out. In the near future, this may provide a portable diagnostic kit; so that the injury can be detected at the point of need. A lab-on-a-chip integrates many laboratory functions on a single chip of only a few mm to a few square cm in size and can handle multiparameter analysis using extremely small fluid volumes of less than pico liters. A lab-on-a-chip is made using silicon, glass, or plastic with tiny pumps and valves, along with sophisticated readout lasers and advanced charge-coupled device (CCD) cameras, etc. A *lab-in-a-phone* is a concept that can convert a standard cell phone into a portable blood diagnosis machine that can detect HIV, malaria, etc.

Nanotechnology supported theranostic agents and targeted drug delivery system are expected to be of great help in providing timely medical help to injured soldiers.

1.4.4.5 In-Situ Tissue Repair

In-situ tissue repair is a part of regenerative medicine in which nanotechnology is applied to stimulate the body to generate new cells *in situ* via bioactive DNA-carrying nanoparticles that induce specific cell growth. Some nanoparticle-based biomimetic scaffolds are developed for optimal uptake and growth of new cells on them. This is further supported by molecular nanomotors to synthesize drugs, repair damaged DNA, and release drugs in a cell.

1.4.4.6 *Artificial Organs*

Many nanotechnology-based approaches are being envisaged for this purpose. Nanomaterials are to be used for selective filtering and purification of the blood (for patients suffering from septic shock and acute poisoning, having renal or liver failure). Scientists have imagined developing a wearable, extracorporeal artificial kidney of nano-dialysis system in combination with a micro-pump and nano/micro-sensors, along with a wearable device for the purification of blood. Smart flexible phones and GPS system for navigation are already on the market to support immediate health care.

1.4.5 Food and Safe Drinking Water

Nanotechnology offers tremendous benefits for improved food processing, packaging and safety and enhances flavor and preserves the nutritional value of foods produced. It is expected to be supported by precision agriculture using sensors (for monitoring soil, crop growth and pathogens), smart pesticides, and the automatic slow release of nanofertilizers as directed by a bio-activated trigger.

Nanofood and supplements, also known as nanoceuticals, containing added nanoparticles are becoming available. Storing degradable vitamins, improving the bioavailability of nutraceuticals and flavor enhancers can be achieved by nanoencapsulating them.

Smart packaging based on nanotechnology is another benefit of nanotechnology in food storage. Nanopackaging material works as a diffusion barrier via antimicrobial and UV-protecting additives that can identify and repel bacteria. In the future there is the possibility of developing a packaging material that can display (based on polymer light-emitting diodes) information on the source, history since production and nutritional status of products, such as self-signaling labels and displays on packaging, by using fluorescent nanoparticles to detect chemical or foodborne pathogens. Use of biodegradable nanosensors for temperature, moisture, time, etc., is also envisaged. Transparent UV-protective films based on TiO_2 nanoparticles along with antimicrobial, antifungal (e.g., silver or zinc nanoparticles) surface coatings are another possible use.

A nanotechnological approach for purifying water by photocatalytic decontamination of water as well as air has been successful through many routes, such as by nanofiltration membranes, attapulgite clay, zeolite, active carbon and polymer filters, using photocatalytic activity of TiO_2 nanosensor for the detection of contaminants. One of the great successes is a portable LifeSaver® water bottle that a soldier can easily carry. This bottle has a membrane having 15 nm pores that prevent bacteria and viruses from passing through.

1.5 Increased Surveillance for Better Protection and Security

Scientists in various defense laboratories are working on a revolutionary idea of manufacturing nano-sized fighter jets that will weigh no more than 10 g. These remote-controlled nano air jets will be as small as a seed and can go anywhere. They can collect military information from both indoors and outdoors and can carry a payload of up to 2 grams. These nanojets will be useful in protecting the lives of soldiers and enhancing operational effectiveness.

Used for its superior strength, CNT can theoretically have tensile strengths of as much as 120 GPa, though actually the highest tensile strength of SWCNT is recorded as 52 GPa and of MWCNT it is 62 GPa. However, the problem is in maintaining its strength. CNT is being envisaged as future weight material for body armor and helmets because it is a very lightweight material.

Protection and surveillance needs keep information in such a way so as to be readable only by those who know how to decode it. The existing cryptosystems use software solutions based on mathematics to distribute keys. The problem is that these mathematics can be copied and decoded and exposed.

Now, a new approach called *quantum cryptography* has been developed. It is an absolute security technology to protect voice, data or video communications using quantum physical phenomena, which makes unbreakable security possible for the first time. Using quantum cryptography the sender transmits a string of bits to the receiver whose values are carried by single photons. Even if an intruder intercepts it, their state gets irreparably changed. The sender and receiver both will detect that it has been intercepted, whereas in digital cryptography, an intruder's interference cannot be detected and messages can be copied.

A single-photon detection device is the ultimate in cryptography communications. These devices make very low-noise, high-speed operation and single-photon generation devices with a low incidence of simultaneous generation of multiple photons. A single-photon generation device contains within a resonator structure, self-organized InAs quantum dots selectively grown on gallium arsenide arsenic, which has confirmed the generation of single photons in the optical communications waveband by electrical drive.

1.6 Smaller, More Effective and Cheaper Nanotechnology-Based Weapons

Defense departments all over the world are trying to develop more lethal weapons than the existing ones. Such weapons focus on precision targeting, minimum weight and signature, optimal impact damage and onboard intelligence. Developing non-lethal weapons is also required, so as not to completely destroy the enemy but rather temporarily neutralize or disable them. Such weapons can be fabricated by using nanotechnology-enabled high strength and lightweight polymer nanocomposites. Stealth and smart skin materials with built-in condition and firing monitoring sensors will also be of great use. Scientists are also trying to develop

(i) *Nano-sized cone materials* that sharpen upon impact and can cause additional damage.
(ii) *Quantum structures* for use as small and efficient directed energy weapons such as directed microwave and laser systems. Nano-dispersed alumina is being used as high energetic propellants. Nanotechnology is also being used to equip the weapon systems with sensors for μ-radar, μ-bolometer (infrared) and acoustic arrays for better targeting.
(iii) *Nanoparticles of aluminium* are used to create compact and powerful bombs. These bombs will create ultra-high burn rate chemical explosives much more powerful than conventional bombs.
(iv) *Mini-nukes* that are very small and light (less than a few kg) are on the agenda of many countries. These mini-nukes use super lasers to trigger comparatively small thermonuclear fusion explosions in a mixture of tritium and deuterium, which will be equivalent to less than a ton to hundreds of tons of high explosives. They are known as future weapons of mass destruction. Since these devices will use very little fissionable material, there will be "virtually no radioactive fallout."
(v) Just like *nano-flotillas* for targeted drug delivery, targeted delivery of highly toxic substances to vulnerable areas can be done using nanotechnology.

1.7 Nanotechnology in Aeronautics for Lighter and Faster Aircraft

Nanotechnology is being envisaged for developing lightweight, intelligence guided, low visual signature, high speed and manoeuvrable fighter jets and missiles; having specific detection and surveillance sensors and weapon systems.

There are many nanomaterials that exhibit ultrahigh strength or mechanical properties. This is especially true for CNTs, which have very high intrinsic strength. Whether used alone or in conjugation with other materials, such as polymers, CNTs offer not only stronger materials but also a substantial reduction in weight. In aircraft as well as land vehicles a lighter material will also be a big support for soldiers. It is expected that nanotechnology will offer an overall weight reduction by a factor of 2–3.

For making lighter and stronger vehicles or aircraft, composites of polymers blended with other nano inorganic chemicals, nanoplatelets or nanotubes in a particular ratio have been developed. The polymers that are usually used are epoxy, PE, PP, ABS, PET, etc., into which nanoparticles are blended to enhance mechanical, electrical or chemical properties. Nanoparticles of different shapes are being used, e.g., spherical, platelets, tubes, fibers, needles, etc. Moreover, the nanoparticles of sepiolite and attapulgite, CNT, layered double hydroxides and smectites, graphene or nanographite platelets, silica, zinc oxide, barium titanate, etc., are also being tried. One such composite that is already on the market is NanoXcel made by Yamaha, which is a nanocomposite of polyurethane matrix in which an exfoliated nanoclay is dispersed. This award-winning composite is lightweight, stiff, strong and durable. Yamaha has used it for hulls and hoods.

To suffice the need of having composites with better electronic and current-carrying properties, *intercalated nanocomposites* are fabricated that have polymer chains alternating with inorganic layers with fixed polymer to inorganic layer ratio.

1.7.1 Exfoliated Nanocomposites

Exfoliated nanocomposites have superior mechanical properties such as (i) high tensile and impact strength, (ii) improved fatigue behavior, (iii) improved flame retardancy, (iv) improved temperature stability, (v) enhanced thermal stability of additives, (vi) reduced linear thermal expansion, (vii) improved crystallinity, (viii) improved resistance to organic solvents, (ix) improved surface finish, (x) enhanced UV-light stability, (xi) reduced oxygen- and water-vapor permeability, and (xii) have new nanopigment basis (PlanoColors®).

Exfoliated nanocomposites have a continuously variable number of polymer chains between the layers that are > 10 nm apart.

1.7.2 Single-Wall Carbon Nanotubes (SWCNT), Double-Wall Carbon Nanotubes (DWCNT) and Multi-Wall Carbon Nanotubes (MWCNT)

These carbon nanomaterials having an inner lumen diameter of 1–2 nm and aspect ratio 103,104 have been found to be good additives for exfoliated nanocomposites. They exhibit up to 1–5 MPa elastic modulus; 30–180 GPa tensile strength; 6000 S/cm electrical conductivity; 2000 W/mK thermal conductivity and a surface area of as high as 1500 m^2/gram. NASA has developed a CNT polymer composite that bends when voltage is applied.

1.7.3 Nanoplatelets and Nanofibers of Graphite/Graphene

Nanoplatelets and graphite/graphene nanofibers are good additives for strengthening nanopolymers. Graphite plates having a thickness of 1–2 nm and an aspect ratio of 103 have induced increased chemical, UV and thermal stability, tensile strength, fracture toughness and diffusion barrier.

Nanofibers synthesized by different methods, such as from vapor-phase grown graphite, electrospun nanofibers or phase-separated liquid crystal fibers, have been found useful in nanofiltration systems. Similarly, nanoclay platelets dispersed in polymer have also been found to be UV-stable and flame retardant.

1.7.4 Electrospun Nanofibers

Electrospun nanofibers are produced by spraying a polymer solution through a fine nozzle by applying 25–50 kV; this reduces the polymer to fiber of 50–200 nm diameters and a few cm long. This process is suitable for producing fibers of ceramic, metal oxide and carbon. Areas in which application of such high throughput nanofibers are being extensively researched are nonwoven mats, nanofiltration units, absorption activity, as catalyst, as reinforcement material for anti-ballistics, and as insulators. Moreover, their selective gas permeability is used to fabricate breathing and protective fabric. Due to the large surface area sensitive to absorption and subsequent change in electrical resistance, they are being investigated for application in sensors.

Another suitable material being considered for lighter and faster aircraft is nanocrystalline metals (< 100 nm) and ceramics. It has been found that by putting the material such as aluminium and titanium under high pressure loads, the internal grain boundary surface area per unit volume increases, thus increasing the hardness and strength may increase by a factor of 2–3.

Helicopters are another consideration. To improve the performance of helicopter rotors, a change in morph per requirement so that it will last longer, as well as reduction in rotor vibration, etc., are being researched.

1.8 Nanotechnology for Stealth Warships and Submarines for Ocean Exploration

Nanotechnology is being used to make lightweight, high strength, smart sensor-guided, low energy consuming, safe, protective and high comfort submarines and naval vessels for ocean exploration. Nanomaterials with such functionalities are nanocomposites of plastics/polymers, metal nanoparticles and carbon nanomaterials, which are foreseen to become future material. One of the necessities for military use is the availability of remote and unmanned guidance, which is becoming possible due to advanced nanosensor and wireless communication.

The power requirement for naval vessels should be controlled by using lightweight ships, energy efficient powering, and reduction of thermal, radar and acoustic signature. For this purpose, the use of electric power systems enabling very low signature, hydrogen fuel cell, miniaturized unmanned vessels are being considered. For operations at sea there is a need to innovate underwater remote, electric and solar-powered ships.

1.8.1 Microwave Absorber for Stealth Technology

Microwave absorbers have the ability to eliminate electromagnetic wave pollution and reduce radar signatures; hence, they are used in civil as well as military applications. As reported by Emerson [3], the first electromagnetic wave absorber came into light in the mid-1930s to improve front-to-back ratio of a 4 GHz antenna. With the development of radar, there has been a growing and widespread interest in radar absorbing material (RAM) technology. RAM is coated on the target whose electrical and magnetic properties have to be altered to give absorption of microwave energy at broad band or discrete frequency [4]. The metallic surfaces of ships are coated with radar-absorbent materials so as to reduce the radar signatures of targets by the reduction of radar cross section, i.e., the actual area "seen" by a radar system when its transmitted wave is reflected from a given object [5]. Due to advancement in wireless and microwave communication technology, a lot of development has taken place in the telecommunication market and the demand for microwave absorbers in the frequency range of 1–20 GHz has increased because of their twofold use as EMI shielding and countermeasure to radar detection. Countermeasure involves the radiation of noise power in the operating band of the radar to confuse or deceive the radar as communication system. Because of these two uses, the suppression of EMI and meeting the electromagnetic compatibility (EMC) have become an essential requirement in industries dealing with high-speed wireless data communication systems operating at 1.9 GHz for handyphone system as well as 2.4 GHz wireless for local area networks (LANs). In addition to mobile communication systems, the satellite communication systems and automatic teller machines (ATM) also operate in the GHz range [6–8].

Microwave absorbers are employed for effective countermeasures against radar surveillance in stealth technology for defense [9]. Application of microwave absorbing coatings on the exterior surfaces of military aircraft and vehicles also helps in avoiding the detection by radar [10].

Nanotechnology-based next generation warships of the United States are in the offing. These warships are all electric. It has been claimed that warships based on this new technology will be more effective, will have longer survivability and will be comparatively cheaper.

1.8.2 Invisible Stealth Ships, Planes and Vehicles

Nanotechnologists have developed invisible left-handed metamaterials that show negative refractive index, and therefore could be virtually invisible to the human eye, roaming radars, and heat-seeking missiles. Although this sounds like a fantasy, it may soon become a reality.

The *left-handed metamaterials* are stacked micro- or nanostructures with resonator capabilities for electromagnetic radiation such as radar, infrared and visible light radiation. At the resonator frequency the metamaterials may show a negative refractive index; when both the permittivity (ε_0) and the permeability (μ) will have negative values. Normal materials have positive values for both the permittivity and permeability and are called *right-handed materials.* In materials with a negative refractive index, incoming radiation does not get reflected or absorbed by the material, rather it is guided and transported along its surface just like in waveguides. Since the radiation is bent around the object, the object cannot be seen but one can see around the object.

Collaborative research by various universities in the USA, Louvain-la-Neuve University in Belgium, Imperial College London in the UK, and Karlsruhe Institute of Technology and University of Stuttgart in Germany have built prototype materials based on this concept.

1.8.3 Radar Absorbing Material: Carbon Nanotubes (CNT)

The first and most common RAM is composed of tiny spheres coated with ferrites. This is painted on the aircraft. Radar waves induce molecular oscillations from the alternating magnetic field in this paint, which converts radar energy into heat. The heat is then absorbed by the aircraft and then gets dissipated.

Another modified RAM that is now used consists of neoprene polymer sheets in which ferrite grains or carbon nanotubes are embedded.

The thickness of the absorbing layer is close to a quarter-wavelength at the frequency of operation, giving a 180° phase difference between the interface reflection and the emergent waves.

1.8.4 Radar Absorbing Material: Ionic Liquids

One more alternative method of microwave absorption being used is poly(ionic liquid)s and ionic liquids containing separated cations and anions which can absorb microwave. Ionic liquid polymers are synthesized by polymerizing monomers containing ionic liquid moieties. These polymers are extremely stable and have high microwave absorption properties. They can be fabricated as films, formulated into coating, painting or other applications.

A Piranha boat using carbon fiber nanocomposite is built by Zyvex Technologies. This boat is 40% stronger and 75% lighter than aluminium. It has shown very economical fuel consumption, hence can stay at sea for a longer time. It is being considered as an unmanned platform for a variety of applications. Moreover, in the future it can replace expensive aircraft carriers.

1.9 Nanotechnology for Vehicles

Vehicles to be used in defense are expected to be lightweight, multipurpose, intelligent guided, energy efficient, safe and protective for passengers. Armor protection and specific monitoring systems for weapons detection and surveillance also have to be taken into account.

A *transformer vehicle* is being developed by the Defense Advanced Research Projects Agency (DARPA) of the United States which can travel on roads, can take off vertically and land. The body of the vehicle could morph itself when airborne (grow wings) or pull them back when it is on land or aloft. It can go over rough terrain, avoid landmines or ambush, while retaining the capability to drive on roads.

1.9.1 Vehicles with Scratch Resistant Surfaces

The scratch resistance of coatings is done on the automotive body and windows (which are mostly made in poly carbonates in modern vehicles) to protect the surface and the underlying layers from mechanical damage, chemical and UV degradation; preferably by incorporating it in the automotive paint systems [11]. Moreover, to increase the fuel efficiency, lighter bodies are made using mold-in-color (MIC) plastics to replace painted interior and exterior components.

To improve the scratch resistance of automotive coats, Barna *et al.* [12] have worked on developing a scratch proof nanocomposite for clear coatings, through the introduction of nanoparticles; because certain nanoparticles can increase the surface hardness and resistance to indentation. The nanoparticles can be uniformly distributed throughout the coatings or they can be designed to preferentially segregate to the top surface.

Another attempt is to make abrasion resistant thin film nanocomposites composed of co-sputtered chrome zirconium (CrZrx) alloys on polymeric substrates. This is due to small grain sizes and complex grain boundaries of crystalline materials and a lower atomic packing density of an amorphous phase [13]. However, many concerns about using nanoparticles are currently being discussed [14] such as the long-term resistance to scratching of the coatings containing nanoparticles primarily at the surface not being known, as the morphological changes during weathering may reduce the effectiveness of this approach.

Another approach to combat scratching is developing it based on self-healing technology.

1.10 Nanotechnology for Satellites

Military satellites are needed for observation, inspection, communication, gathering information, etc. The advantages of swarms of nanosatellites for defense are that they can be used for destruction of other spy satellites; and can intercept communications or make observations.

Nanotechnology is being applied to make lightweight, miniaturized components and hence very small satellites. This is expected to not only make them cheaper to manufacture, but also to launch them into orbit in space. Trials are ongoing for a distributed network of small satellites instead of one large one. Based on their weight these will be minisatellites (50–500 kg), microsatellites (10–50 kg), nanosatellites (1–10 kg) and even beyond nano, picosatellites (< 1 kg). Each micro-satellite will have its own solar panels as a power source. The ultimate goal is to develop a complete satellite-on-a-chip, the so-called picosatellites. One Delfi-C3 microsatellite was already launched in 2008 for a 3 kg CubeSat measuring 10×10×30 cm. It flies thin-film solar cells, two autonomous wireless sun sensors and a miniaturized UHF-VHF transponder.

1.11 Nanomaterials for Portable Energy/Power

Energy supply is needed for all the combat systems in defense, such as for wearable invisible or camouflage suits or garments, helmets or head protection and hand weapons, integrated sensors, communication, etc. The energy supply systems have to be low weight miniature power sources because there is a demand for lightweight batteries with a high energy density for wearable electronics, electric vehicles and robotics (Table 1.1).

The areas in which efforts are being made are described below.

1.11.1 Portable Fuel Cells (FC)

Fuel cells are composed of nanoporous membranes and nanostructured catalysts. Available portable fuel cells are solid oxide fuel cells running on 2.5 kg propane and hydrogen fuel cells.

Table 1.1 Different battery technologies used in vehicles and their energy density.

Battery technology	Energy density (Wh/kg)
Old: Lead-acid	30–40
Mature: Nickel Metal Hydride	30–80
State of the art: Li-ion	150–300
-Li carbon/iron-phosphate	
-Li vanadium oxide/cobalt oxide	
-Li polymer	
In development: Na-ion	400
Future: nano Li-ion	1000–2000
- Li-vanadium oxide aerogel	
- Silicon nanowire Li-ion	

1.11.2 Rechargeable Lithium (Li) Batteries

A Li-ion battery stores electrical energy in a chemical form. Nanoparticles are being developed for metal oxide nanocomposite cathodes because they offer higher density lithium intercalation, excellent electrical conductivity and improved diffusion. For the anode, graphite and metal alloys are used. It offers higher tensile strength to cope with the increase in volume changes during the charging cycle and provides 3x higher energy electrode material.

1.11.3 Supercapacitor

Supercapacitor stores electrical energy directly as charge on sets of electrodes that are separated by an insulator and covered with a thin coating of electrolyte. Since the electrode surface area determines the power density in a supercapacitor, CNTs, nanocrystalline materials, and aerogels are being developed because they have a large surface-to-volume ratio. So far, the developed systems offer capacitive energy storage of up to 200 F/gram.

1.11.4 Solar Cells

Nowadays, 90% of solar cells are made of silicon. Silicon wafers are costly to manufacture and have an efficiency of 14%. However, up to 50% efficiency has been achieved in a lab with multilayered III-V semiconductors (GaAs/InSe/GeTe). These are even more costly than the solar cells that are used in the space satellite systems. To reduce the cost and increase the efficiency, nanotechnology-based solar cells are being developed, such as

- ***Silicon Nanowire Solar Cell*** — This type of cell has not shown very high efficiency, with only a maximum efficiency of 18% so far.
- ***Nanowire Solar Cell*** — This is another nanowire solar cell which is based on indium phosphide (InP) nanowires on the metal electrode of indium tin oxide (ITO); and then the nanowire-electrode platform is covered in the organic polymer, poly(3-hexylthiophene). Once the electron and hole split, the electron travels down the nanowire - the electron superhighway - and merges with the electron-capturing electrode. This rapid shuttling of electrons from the p-n

junction to the electrode may be a more efficient future photovoltaic device.

- ***Organic Dye-Sensitized Solar Cell (DSSC)*** — This cell uses a dye molecule to absorb a photon and TiO_2 nanoparticles to capture the electrons. The cost of DSSC is 60% less than the silicon-based solar cells; however, its efficiency is hardly 10%.
- ***Quantum Dot Solar Cell (QDSC)*** — This type of cell is expected to give as high as 86.5% efficiency. In QDSC the absorption can be tuned to any wavelength of visible light by altering its size. Hence, there is a possibility of generating three electrons per photon, as compared to only one for existing silicon-based technologies.

1.12 Nanosensors

It can be noted that almost all developments and research work related to defense applications use sensors in one form or the other; for example, sensing light by photocells, sound by microphones, ground vibrations by seismometers and force using accelerometers. Moreover, there are sensors for sensing magnetic fields, electric fields, radiation, strain, acidity, etc. In our everyday life we also depend on sensors such as metal detectors at airports, malls, hotels, etc., and smoke detectors for building protection. Computers have greatly benefited from the advent of nanotechnology, which has revolutionized computers. Different types of ultrasensitive nano-sized sensors are now being developed for various applications. The use of nanotechnology for developing sensors is focused on (a) *Smaller and ultra-compact sensors* with the help of micro-electromechanical systems (MEMS); these nanosensors have an electronic nose and cantilever molecular sensor. (b) *Smarter devices* are another new mantra of nanotechnology. The nanosensors have built-in "intelligence" which can store and process data on the spot, selecting only the most relevant and critical items to report. (c) *Faster mobility* with the help of today's wireless networking technologies; such sensors can send back data from remote locations or even while they are in traveling mode. In one integrated chip these nanosensors have advantages of function integration by sensing, data processing, storage, wireless communication in a network, multi-parallel analysis for high throughput, matrix arrays for high zeptogram (a zeptogram is 10^{-21} g) sensitivity and directional information. Their production uses less chemical waste, is low cost and has low power consumption. They are capable of of energy scavenging (solar, temperature, mechanical) for continuous power supply. Being very small, they are portable, wearable, point of analysis and self-operating in remote locations and are disposable.

Many sensors have already been developed taking advantage of the unique properties of nanomaterials to become smaller and more sensitive compared to conventional technology. Portable, efficient and highly sensitive infrared thermal sensors will be a very valuable addition to the military such as small, lightweight accelerometers, GPS for motion and position sensing, health monitoring sensors and drug/nutrition delivery systems. Robotic drones are already operated by remote control.

1.12.1 Chemical Nanosensors

Chemical nanosensors are built by using CNT for sensing properties of gaseous molecules and ionization of gaseous molecules, while titanium nanotubes are used to detect atmospheric concentrations of hydrogen at the molecular level.

1.12.2 Mechanical Nanosensors

Mechanical nanosensors work by displacing a nanocantilever, beam or nanowire under the influence of an inertial force, vibration or pressure difference. A change in a system can be measured by capacitive displacement sensors or electronically by using field effect transistors, optically by laser deflection or via surface charge on piezoelectric potential of zinc oxide (ZnO). Forces as small as 10–18 N (N is Newton symbol) 1 $kg \cdot m \cdot s^{-2}$ can be detected. The mechanical sensors currently in use are 3D acceleration sensors, pressure sensors and vibration sensors.

1.12.3 Magnetic Nanosensors

Magnetic nanosensors are fabricated using unique magnetic properties of very small nanoparticles that exhibit quantum properties. These ultrasensitive sensors have a sensitivity of 10^{-9} Tesla.

1.12.4 Radiation Nanosensors

Such sensors are used for sensing electromagnetic radiation. For this purpose, a nano-sized dipole antenna with optical dimensions (50 nm – 100 μm length) connected to a microbolometer matrix array is used. A more sensitive radiation sensor is made by using a quantum well structure. In quantum wells, electrons tunnel through a barrier when activated by external radiation. This has a high signal-to-noise ratio and the presence of many quantum wells on a chip yields a large signal output. It has found applications in infrared and THz radiation.

1.12.5 Portable Miniature X-Ray Nanosensors

These portable nanosensors are fabricated using carbon nanotubes (CNTs). Because CNTs are very effective electron emitters, they are used as electron source to generate X-rays used in portable X-ray detection devices.

1.12.6 Surface-Enhanced Raman Spectroscopy (SERS) Nanosensors

The SERS nanosensors are used for detection of up to parts-per-trillion range of explosives and contraband molecules such as TNT, DNT, RDX, TATP, PETN, DMDNB and cocaine.

1.12.7 Smart Dust Sensors

Smart dust sensors are a very tiny network of sensors that detects data about light, temperatures or vibrations and then transmits it to larger computer systems. However, this technology has not yet reached reality.

1.13 Nanotechnology for Logistics

Technologies for logistics are concerned with how to perfect the security and safety of transported goods and how to improve the speed and efficiency of the logistics chain. This needs automated, unmanned operations for tracking, tracing and identification of security mishaps. Nanotechnology and advanced robotics operations deliver critical payloads and materials to designated sites that may be miles across control system. There will be a need for unmanned ground and aerial vehicles linked to high-precision, multi-platform, and strike capabilities for protecting supplies in transit.

Needless to say, nanotechnology can offer logistics lightweight containers made of nanocomposite polymers that will also be able to absorb shock, reducing the impact of G-forces.

Self-signaling containers made up of smart materials and sensors (chemical, vibrations, heat) to alarm for any deficiencies in display functions are being developed. It is believed that the nano-assemblers will reduce logistics.

1.13.1 Smaller, Faster Nano-Cameras

In 2013, a group of scientists from MIT media Lab (Ramesh Raskar, Kadambi, Refael Whyte, Ayush Bhandari, and Christopher Barsi) and the University of Waikato in New Zealand (Adrian Dorrington and Lee Streeter) were in the news for producing an inexpensive "nano-camera" that can operate at the speed of light. Apart from the use of this device in medical imaging, collision-avoidance detectors for cars, and interactive gaming, its very critical role in defense has also been envisaged. This three-dimensional camera has found use in improving the accuracy of motion tracking and gesture-recognition devices. The camera is based on "time-of-flight" technology, in which the location of objects is calculated by how long it takes a light signal to reflect off a surface and return to the sensor. It can operate in rain, fog and even through a translucent object, so it cannot capture translucent objects in 3D. That is because the light that bounces off the transparent object and the background smear into one pixel on the camera. But according to the inventors, translucent or near-transparent objects can be generated using these 3D models. The "nano-camera" can probe the scene with a continuous-wave signal that oscillates at nanosecond periods; and using light-emitting diodes (LEDs) can strobe at nanosecond periods, i.e., the camera can reach a time resolution within one order of magnitude of femto-photography. Just by changing the code, light paths are unmixed and therefore light moving across the scene is visualized. One of the accomplishments of NASA was to develop a new class of camera using nanotechnology where a photon-plasmon-electron conversion approach demonstrated a new detector imaging that converted photons from an incoming optical signal into surface plasmon waves. The plasmon waves are formed in a 100 nm thick metal film that is bonded to a semiconductor substrate. A slit in the metal film helps to collect plasmon waves, where their energy is focused to produce electron/hole pairs in the semiconductor substrate. The electron/hole pairs generate a measurable electrical signal. Slit orientation is very important here because it helps in producing different polarization signals from the same imaging chip. The imaging chip has different slit sizes and orientation can produce multi-spectral signals.

1.14 Conclusions

For application in military and defense, Nanoscience, Nanoengineering and Nanotechnology will play a significant role. Fundamental knowledge of nanoscale science and engineering is a must for optimization of any process. This chapter presented a brief overview of defense-oriented nanotechnology-based developments. There are many concepts and theoretical possibilities that are still in R&D stages. However, many weapons, structural material, sensors, protective materials, electronics, etc., are already in military use. The progress in nanotechnology in biology and medicine, energy and power generation, information and communication technology has been tremendous. Significant investments are being made worldwide on nanotechnology-based research for military uses. Many nanoproducts already exist and are commercially produced, such as:

- Accelerometers
- Additives in food and cosmetics
- Condition sensors
- Filters
- Lab-on-a-chip systems
- Nanocatalysts
- Nanocomposites (nanoclay and carbon nanotube composites)

- Paints
- Polymers
- Tire pressure sensors
- Wireless sensor networks

The defense organization is gaining a lot of benefit from these nanoproducts and is now incorporating them in defense-related devices. In the future, nanotechnology can offer soldiers better equipment, higher security and safety and effective operational capacities. It will also provide better warships, better advanced weapons and vehicles.

References

1. Ni, X., Wong, Z.J., Mrejen, M., Wang, Y., Zhang, X., *Science*, 349, 6254, 1310–1314, 2015, DOI: 10.1126/sci.ence.aac9411.
2. La Spada, L., McManus, T.M., Dyke, A., Haq, S., Zhang, L., Cheng, Q., Hao, Y., *Sci. Rep.*, 6, 29363, 2016, DOI: 10.1038/srep29363.
3. Emerson, W.H., *IEEE Trans. Antennas Propag.*, 21, 4, 484–490, 1973.
4. Meshram, M.R., Agrawal, N.K., Sinha, B., Misra, P.S., *J. Magn. Magn. Mater.*, 271, 2-3, 207–214, 2004.
5. Pinho, M., Silveira, P., Gregori, M.L., *Eur. Polym. J.*, 38, 2321–2327, 2002.
6. Ramasamy, D., *Proceedings of INCEMIC 97, 7B-7, IEEE Conference*, New Jersey, pp. 459–466, 1997.
7. Chung, D.D.L., *Carbon*, 39, 279–285, 2001.
8. Horvath, M.P., *J. Magn. Magn. Mater.*, 215–216, 171–183, 2000.
9. John, D. and Washington, M., *Aviat. Week Space Technol.*, 129, 28–29, 1998.
10. Stonier, R.A., Stealth aircraft and Technology from World War II - the Gulf. *SAMPE*, 27, 4, 9–17, 1991.
11. Sharon, M. and Sharon, M., *Nano Dig.*, 2, 2, 16–20, 2010.

12. Barna, E., Bommer, B., Kursteiner, J., Vital, A., Trzebiatowski, O., Koch, W., Schmid, B., Graule, T., *Compos. Part A*, 36, 473–480, 2005.
13. Evans, D., Zuber, K., Murphy, P., *Surf. Coat.*, 206, 3733–3738, 2012.
14. Shi, X. and Croll, S.G., *J. Coat. Techol. Res.*, 7, 73–84, 2010.

2

Stealth, Counter Stealth and Nanotechnology

Madhuri Sharon

Walchand Centre for Research in Nanotechnology & Bionanotechnology, WCAS, W.H. Marg, Ashok-Chowk, Solapur, Maharashtra, India

Nanotech-enabled military hardware – from the combat soldier to ultra complex space-based nanobot arsenals – will make any current Weapons of Mass Destruction seem as primitive as the sticks and stones used by low-tech Neanderthal warriors.

Chris Phoenix
Director of Research
Center for Responsible Nanotechnology

2.1 Introduction

This chapter presents certain emerging aspects of stealth and counter-stealth technologies and the latest efforts in advancing the use of nanotechnology in them. Stealth and counter-stealth technologies are part of defense security. Defense demands both offensive and defensive approaches for safety. The latest approach for air safety relies on stealth vehicles, high technology

Madhuri Sharon, et al. Nanotechnology in the Defense Industry: Advances, Innovation, and Practical Applications, (37–88) © 2019 Scrivener Publishing LLC

and high-definition active and passive sensors, and also electromagnetic-based anti-stealth systems.

This chapter looks into both the offensive and defensive dimension of air defense. Protection of aircraft involves developing stealth, anti-stealth, high-definition sensors, energy efficient solar-power-driven aircraft, etc. Stealth technology is not a single technology. It is a combination of technologies that attempt to greatly reduce the distances at which a person or vehicle can be detected; in particular, radar cross-section reductions, but also acoustic, thermal, and other aspects.

Prior to discussing the importance of nanotechnology in stealth, it is important that we first understand what stealth is and why it was developed. For this we have to look into radar, which provided an incentive for developing stealth. Moreover, we will also look at the supporting facilities incorporated in stealth that are based on nanotechnology.

2.2 Radar – An Incentive for Developing Stealth

2.2.1 Principle of Radar

Radar is a system used to detect the position of a moving object, especially airplanes. The principle or concept of radar is based on two understandings:

(i) **Echo** can be considered as a wave bouncing off the surface and coming back to the source, which makes it possible to predict the radar signature an aircraft made with flat panels, called facets. This principle can be used to detect time and distance of the target. A simpler way of describing an *echo* is that when a sound is made, sound waves bounce off of a surface (either the water at the bottom of a well, a canyon wall that is far away, or from the hills of a valley) and travels back to our ears. The length of time between the origin of the sound and the moment that you hear the echo is determined by the distance between you and the surface that creates the echo. The echo of a sound can be used to determine how far away something is.

(ii) **Doppler shift** is the second principle of radar, which is used to detect the speed of an approaching target. Doppler shift occurs when sound is generated by, or reflected off a moving object. Doppler shift in the extreme creates sonic booms. To better understand Doppler shift imagine a horn blaring in a vehicle approaching you at 100 km/h; you will hear a particular "note" of the horn, but as the car passes by you, there will be a shift to a lower note, though the horn is making the same sound all the time. The change you hear is caused by Doppler shift. The Doppler shift of the echo can determine speed of the object.

2.2.2 How Radar Functions

Radar transmits a certain amount of electromagnetic energy through a directional antenna. This energy focuses it into a conical beam. When a reflective target blocks part of the beam, that beam is reflected in different directions, and some energy that is reflected in the direction of the radar antenna is detected (Figure 2.1). Radar transmits this energy in thousands of pulses per second. In the gaps between the pulse transmissions, the radar becomes a receiver. The time interval between the transmission and reception of the pulse gives the range from the radar to the target. Radar does not see

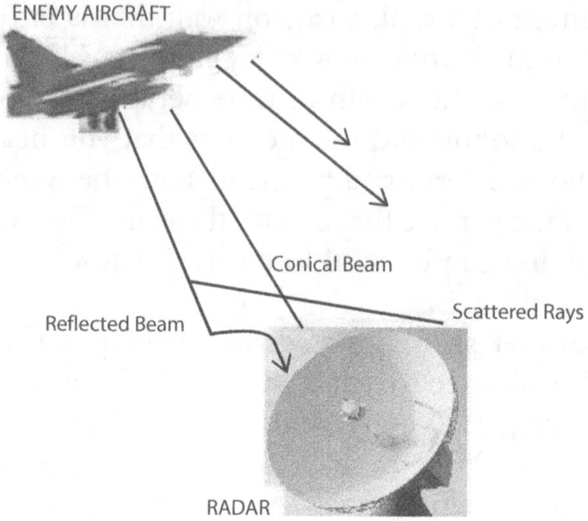

Figure 2.1 Schematic diagram showing how a signal is transmitted and received in a Radar system.

visible light, it only sees the energy that is reflected towards it and when its antenna captures enough energy to rise above the electronic noise that is present in the receiver.

2.3 What Is Stealth and Why Was It Developed?

The first radar tracking systems were developed in the late 1930s to detect the presence of aircraft in a particular vicinity. This was seen as a threat to the military. Therefore, to protect military aircraft from being detected by the enemy, the need to develop stealth technology arose. Aircraft was designed to make them almost invisible to radar detection and also to sonar and infrared (IR) detection methods (Figure 2.2). Most of the latest advanced stealth aircraft are: Northrop Grumman B-2 Spirit (also known as a stealth bomber), Lockheed Martin F-22 Raptor (a fifth-generation, single-seat

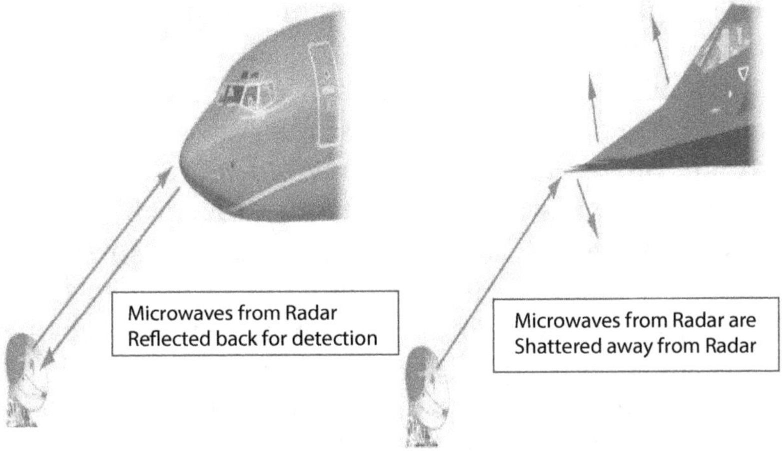

Figure 2.2 Alteration in design of stealth aircraft to avoid easy detection by radar.

stealth), Lockheed F-117A Nighthawk, Sikorsky UH-60 Black Hawk (stealth helicopter), SU-35, Mikoyan MiG-35, Chengdu JXX/J-20, Advanced Medium Combat Aircraft (AMCA), etc.

The first two stealth aircraft produced were used in combat in 1989 in Panama and the other were the F-117s used in 1991 in Iraq.

Stealth materials used to reduce radar reflectivity, ultraviolet/infrared signatures and acoustic signatures include specific paint coatings such as (i) Conductive carbon, metals and glass fibers; conductive inks or paints with small cell foams; (ii) fine grained ferromagnetic or ferrite particles suspended in a variety of rubber, paint, or plastic resin binders; (iii) ceramic coatings heavily loaded with electrically conductive fillers or ferromagnetic particles; (iv) radar absorbing honeycomb material that is very lightweight composite with open cells; (v) transparent radar absorbing material; (vi) alteration in the design to avoid detection by radar and (vii) infrared treatments by paints and coatings.

Making an aircraft stealth demands complex design philosophy, as it has to work on several fronts such as reducing the ability of an opponent's sensors to detect, track and attack. Making an aircraft almost invisible makes it stealth. This technology has been developed to camouflage the military aircraft so that they are difficult to see, whether on the ground or in the air. Since the main goal of stealth technology is to make an airplane invisible to radar, two different ways were created to achieve invisibility; one is to shape the airplane in such a way that any radar signals it reflects are reflected away from the radar equipment; and the other is to coat the airplane with materials that absorb radar signals. However, there are six other disciplines in which an aircraft has to be modified to be stealthy, i.e., radar, infrared, visual,

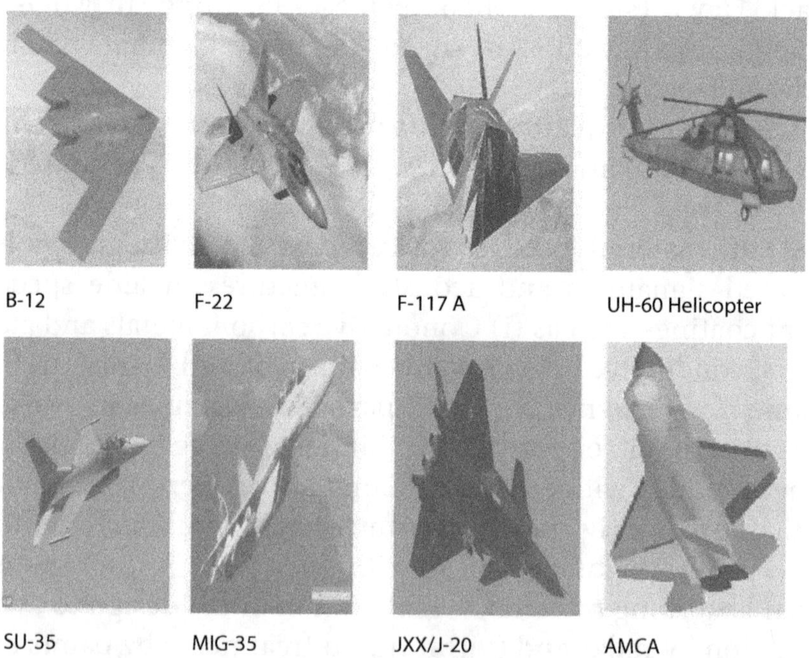

Figure 2.3 Some popular stealth aircraft used in defense.

acoustic, smoke and contrail. Some of the popular aircraft designs in use are shown in Figure 2.3.

2.4 Considerations and Efforts for Designing Stealth Aircraft

Almost since the invention of radar, the following various methods have been tried to minimize detection.

2.4.1 Camouflaging

Camouflaging was the first and simplest attempt to make a plane less visible to the enemy. Many camouflage schemes have been tried such as:

(i) **Counter shading** the aircraft is done to counter visibility, i.e., coating the upper surface with a disruptive pattern of ground colors (green and brown), lower surface with sky color, and aircraft that fly by night with black color. Generally, stealth aircraft use matte paint.

(ii) **Active or adaptive camouflage** is another technology that allows colors, patterns and brightness to rapidly change to match the background or the surroundings. The idea of active camouflage is copied from several animals (reptiles, cephalopods such as molluscs, squid, and flatfish of the sea) that are protected by changing their color. This technology was developed during World War II. Nowadays, gray paint coating in disruptive schemes and Yehudi lights are also used for active camouflaging.

(iii) **High flying aircraft** with side-looking airborne radar (SLAR), forward-looking infrared (FLIR) cameras, etc., are also used.

2.4.2 Plasma Active Stealth

Plasma is a gaseous medium in which atoms are broken up into free-floating negative electrons and positive ions. Ions are atoms which have lost electrons and are left with a positive electric charge. Plasma is sometimes called "the fourth state of matter," beyond the familiar three states, e.g.

SOLID	LIQUID	GAS	PLASMA
[ICE]	[WATER]	[VAPOR]	[IONIZED GAS]

In plasma stealth, an ionized gas "layer" is generated, which surrounds and envelops the aircraft making it invisible (Figure 2.4, right) and reduces the radar cross section (RCS). It is a quasi-active system in which radar signals are received and absorbed/scattered by plasma that is capable of absorbing a wide range of radar frequencies, angles, polarizations, and power densities. The wavelengths absorbed by plasma are converted to thermal energy rather than getting reflected.

Stealth technology reduces the detection range of an aircraft by making the aircraft partially invisible to radar or any other

Figure 2.4 (left) Plasma-stealth aircraft and (right) stealth surrounded by plasma.

means of detection. Stealth technology does not make an aircraft completely invisible to radar.

2.4.3 Inactivating Radar Signal or Making Planes Less Visible

To nullify the activity of radar, the use of a RAM (radar absorbing material) coating was envisaged. Radar absorbing material or microwave absorbing material is the most recent technology aimed at minimizing the aircraft's radar cross section (RCS) and infrared signature. By using RAM, the RCS is reduced because decreasing RCS reduces the radar signature.

The Horten Ho 229 flying wing jet was the earliest known stealth aircraft that used carbon powder in the glue to absorb radio waves. With the development of radar, there has been a growing and widespread interest in RAM technology that uses microwaves, as it is necessary to develop stealth that can absorb microwaves. Nowadays, microwave absorbers are used both in civil as well as military applications. Apart from being used in aircraft to make it stealth, a recent development has opened the door for the use of microwaves in telecommunication technology also. The demand for various kinds of microwave absorbers has increased in the frequency range of 1–20 GHz because of their twofold use as EMI shielding and as a countermeasure to radar detection (countermeasure involves the radiation of noise power in the operating band of the radar to confuse or deceive the radar as communication system). Out of these twofold uses, the suppression of EMI and meeting the Electromagnetic Compatibility (EMC) have become essential requirements in industries dealing with high-speed wireless data communication systems operating at 1.9 GHz for personal handy-phone system (PHS) and 2.4 GHz wireless for local area networks (LANs). These kinds of systems are also employed in mobile communication fields such as in the electronic toll collection (ETC) system which operates at 5.8 GHz, where it prevents incorrect operation [1]. In addition to mobile communication systems, the satellite communication systems and automatic teller machines (ATM) also operate in the GHz range [2–4]. Therefore, in order to allow the coexistence of all those various systems and instruments without harmful electromagnetic interferences, there is a need to develop new shielding and absorbing materials which especially require anti-electromagnetic interference coatings, self-concealing technology and microwave darkrooms [5–7] with high performance and a large operating frequency band [8]. The application of the appropriate microwave absorbing material in electronic equipment helps to control the excessive self-emission of electromagnetic waves and ensures undisturbed functioning of the equipment in the presence of external electromagnetic interference.

Stealth technology for defense is an area where these microwave absorbers are employed for effective countermeasures against radar surveillance [9, 10]. Application of microwave absorbing coatings on the exterior surfaces of military aircraft and vehicles also helps in avoiding detection by radar. This technology is also applied to unmanned combat aerial vehicles (UCAV) such as the Boeing Phantom Ray.

2.4.3.1 Radar Absorbing Material (RAM)

Microwave history since 1930 has shown that there has been a continual need for developing efficient microwave absorber. With the increasing use of microwave in various fields, the need for absorber was developed because of its countermeasure. RAM is based on a principle similar to that of reflecting a light beam from a mirror. Placement of light source at an angle at which the reflection takes place decides the amount of light that gets reflected; e.g., if a mirror is rotated from 0 degrees to 90 degrees, at 90 degrees the maximum amount of light is reflected back to the same direction of the light source. On the other hand, when the mirror is tilted above 90 degrees, as it

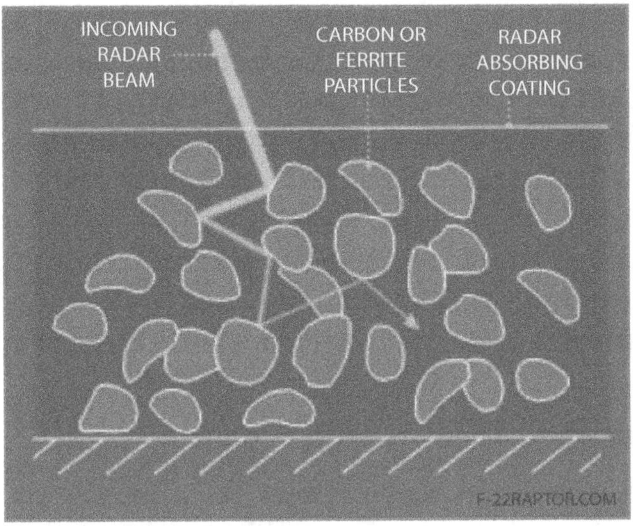

Figure 2.5 Ferrite acts as radar absorbing material.

proceeds to 180 degrees the amount of light reflected in the same direction decreases drastically.

To make an aircraft stealth its surface is coated with paint that is deposited on the surface like colonies of tiny pyramids and the gaps between pyramids are filled with ferrite-based RAM (Figure 2.5). The waves hitting various angles of these tiny pyramids deflect the incident radar energy in different paths or directions, thus creating a maze of RAM. So, the pyramidal structure acts as RAM and behaves like a filter as well as absorber of microwave or electromagnetic energy. Let us now find out what microwaves are.

2.4.3.2 What Are Microwaves?

Microwaves are electromagnetic waves with frequencies between 300 MHz (0.3 GHz) and 300 GHz in the electromagnetic spectrum. As the name implies, radar absorbing materials or RAM are coated on the target whose electrical and magnetic properties have to be altered in order to absorb microwave energy at broadband or discrete frequency. The use of radar absorbing materials to coat the metallic surface of ships is one method of reducing the radar signatures of targets by the reduction of radar cross section, i.e., the actual area "seen" by a radar system when its transmitted wave is reflected from a given object.

On account of their ability to eliminate electromagnetic wave pollution and to reduce radar signatures, microwaves are used in stealth. There are various other applications of microwaves.

2.4.3.3 How Are Microwaves Absorbed?

The properties of microwaves vary according to the medium through which they propagate. Therefore, there is a need to know the propagation of electromagnetic waves through different material medium; especially the dielectric and conducting materials (e.g., ferrites, metal particles, carbon or complex materials) which have both dielectric and magnetic loses. In the proceeding paragraphs, the theory of propagation and absorption of microwaves through different media and corresponding boundary conditions are discussed in brief, because microwaves show the variation in their properties when they are propagated through different medium. The propagation characteristics of electromagnetic waves depend on the electrical parameters (σ, ε, μ) of the medium and the presence of the boundaries or interfaces between the two media. In free space, at sufficiently large distance from the source, wave propagation takes place in transverse electromagnetic mode (TEM) from one point to another, where E and H fields are orthogonal to each other and both are perpendicular to the direction of propagation for sinusoidally time varying vector fields, as shown in Figure 2.6 below. The two fields, i.e., E and H, are related by impedance, called the Wave Impedance (η), which is

$$\eta = |Et| / |Ht| \tag{2.1}$$

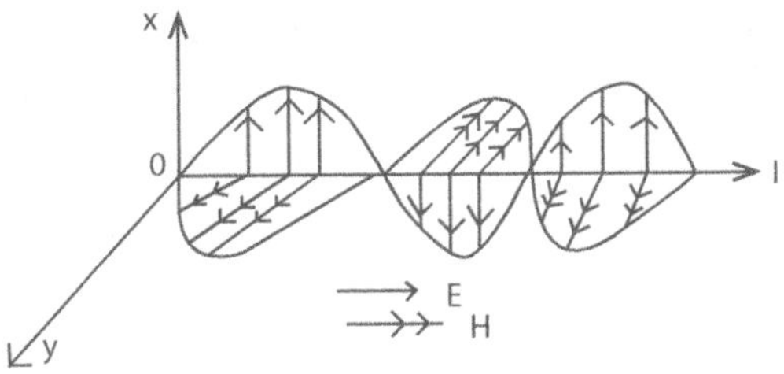

Figure 2.6 Transverse electromagnetic mode wave propagation.

Here, Et and Ht represent transverse components of electric and magnetic fields with respect to the direction of propagation.

By solving Maxwell's equations involving E or H fields, we get

$$\nabla^2 E + K^2 E = j\omega\mu J - \nabla \nabla . J / j\omega\varepsilon \; ; \; \nabla^2 H + K^2 H = -\nabla . J \tag{2.2}$$

$$\text{Where, } K = \sqrt{-j\omega\mu(\sigma + j\omega\varepsilon)} = \sqrt{-j\omega\mu\sigma + \omega^2 \mu\varepsilon} = \alpha + j\beta \tag{2.3}$$

Here, **K= α + jβ** is called the propagation constant of the medium, where α is the real part of Propagation Constant measured in decibels or nepers/meter (1 neper = 8.686 decibels) and β is phase constant, which is the imaginary part of propagation constant measured in radians per meter. The real part and imaginary part of the propagation constant give an idea about whether the loss of electromagnetic wave is by attenuation or change of phase or by both, in the medium. Therefore, it is important to understand the nature of K when the electromagnetic wave propagates in the different medium, for example:

*(i) **Medium with finite conductivity***

When electromagnetic wave propagates in a medium with finite conductivity σ, a conduction current density

J = σE exists, which gives attenuation α because of joule heating.

In the case of electromagnetic wave at frequencies below the optical region (e.g., microwave region) the propagation constant in a good conductor (σ >>ωε) becomes,

$$K = (1 + j)\sqrt{\pi f \mu \sigma} = \alpha + j\beta$$

(ii) A non-conducting or lossless medium

In a non-conducting or lossless medium,

$$\mathbf{J = 0 \text{ as } \sigma = 0}$$

For such medium the propagation constant is

$$K = \omega\ (\mu\varepsilon)\ 1/2 \tag{2.4}$$

(iii) A free space

In free space where (**σ ~ = 0**),

the propagation constant is **K= ω (μ0 ε0)1/2**

where, μ0 and ε0 are the permeability and permittivity of free space respectively.

When an electromagnetic wave propagates in a medium, it exhibits one important characteristic, which is that it can transport energy from one point to another point. The time average power transmitted through a unit area is given by,

$$\mathbf{P} = 1/\mu 0\ (\mathbf{E} \times \mathbf{B}) \text{ or } (\mathbf{E} \times \mathbf{H}) \tag{2.5}$$

Where, **P** is Poynting vector, **E** is the electric field, **H** and **B** are the magnetic field and magnetic flux density respectively (bold letters indicate vector quantities), and μ is the permeability of the surrounding medium.

For an electromagnetic wave, propagating in free space μ becomes μ0, the permeability of free space. The Poynting vector is derived by considering the conservation of energy and taking into account that the magnetic field cannot do work. This vector especially plays an important role in the explanation of loss of electric field from the electromagnetic wave in a material medium.

Another consideration is loss of microwave in different conditions. As mentioned above, microwave is an electromagnetic wave below the optical region; when these waves propagate in a medium, the impedance of the wave changes with the material medium in which it propagates. The propagation constant (i.e. sum of attenuation and phase constant) for the electromagnetic wave in different medium changes with the wave impedance, which gives an idea about the loss of electromagnetic wave in that medium.

2.4.3.4 Microwave Transmitting Structures

It is important to understand the microwave transmitting structure, such as rectangular wave guide and planar transmission line, especially the microstrip line. The microwave circuits and devices constitute a section or sections of microwave transmission lines or waveguides. Microwave signals are propagated through these lines as electromagnetic waves and scattered from the associated junction of these lines to travel in well-defined direction or ports.

Conventional open wire lines are unsuitable for microwave transmission because of the high radiation losses that occur when the wavelength becomes smaller than the physical lengths of these lines at high frequencies.

The structures considered by David Pozar [11] are:

1. *Single conductor lines* such as rectangular, circular and ridge waveguides. In this category of lines, the mode of transmission is a TEM or quasi-TEM wave.
2. *Multiconductor lines* such as coaxial lines, strip lines, microstrip line, slot lines and coplanar lines. Here the modes are either TE or TM waves or both.
3. *Open boundary structures* such as dielectric rods. These lines support a combination of TE and TM waves called hybrid HE modes, except possible axis-symmetric modes, which are either purely TE or TM waves.

The relation for input impendence, reflection coefficient, transmission coefficient, characteristic impendence, all are applicable to microwave guides operating in a single mode [12].

2.4.3.5 *Types of Microwave Absorbers*

Here we discuss the microwave system used for the absorption study. Various types of microwave absorbers and their loss mechanism are introduced and the concept of metal absorber is discussed with its loss mechanism. This section ends with the limitations of the available absorber. Moreover, it also deals with the need for nano-absorber and ferrite nano-absorber and its limitations.

From the above discussion it is seen that absorber material should have dielectric, magnetic and conductive losses because the manner by which these properties change with frequency reflects the performance of the absorber [13, 14]. The absorbers work by allowing the wave to penetrate into a region where the electric and/or magnetic fields experience loss. All lossy materials are frequency dependent. The frequency dependence constrains the values of real and imaginary parts of the relative permittivity (**εr**) and relative permeability (**μr**) of the material. Conductivity is another loss mechanism; however, it also affects the material's impedance, because

$$\varepsilon = \sigma/\omega\varepsilon 0.$$

Thus, a metal (σ =107 mhos/m) is both highly lossy if it is used as conductive filler and highly reflective if its thickness is greater than the skin depth. Therefore, based on the principle of absorption the microwave absorbers are categorized as *resonant absorber, broad-band absorber, magnetic absorber* and *dielectric absorber*. These four types of absorbers are discussed in brief below.

2.4.3.5.1 Resonant Absorber

The resonant absorbers are much thinner than the other types; they are preferred for most applications where wide operational bandwidth is not a requirement. They are customarily chosen to improve antenna patterns by covering parts of antennas and to improve radar performance by covering nearby reflecting surfaces. Their physical flexibility is often useful in coating surfaces of complex shape. Usually, these absorbers are directly mounted on the metal, causing the reflection. Even the surface is sharply curved or small in terms of wavelengths; some reduction in performance is to be expected with the application of this absorber. For cases where the material cannot be mounted directly on metal, it is necessary to provide a conducting rear surface, such as foil or silver paint, to achieve low reflection at the resonant frequency. In resonant absorber a wave incident upon the surface of the layer of dielectric material is partially reflected and partially transmitted. The transmitted portion undergoes multiple internal reflections to give rise to a series of emergent waves, thus causing the loss. The limited reflectivity characteristics of resonant absorbers lead to the use of the thicker broadband types in many applications.

2.4.3.5.2 Broadband Absorber

In contrast to resonant absorber, which uses the back-face reflection, the broadband absorbers need sufficient attenuation so that any energy reflected from the back face should be able to leave the material with a very low-level reflection. To get this degree of attenuation, the absorber should have appreciable thickness in terms of wave length. Hence, a material with higher loss tangent and higher dielectric constant that can provide adequate loss at reduced thickness is used. It is important to note that a medium with such properties would also have impedance quite different from that of free space, and therefore will show a higher value in front-face reflection. Such front-face reflection can be significant in a lossy medium and it strongly deteriorates the performance of the absorber.

To understand the principle of broadband absorber, one must understand the concept of tapered impedance. These absorbers are designed by varying the impedance from that of free space at its incident surface to a low impedance, lossy material at its rear (back) surface. When microwave incident energy is absorbed on such absorbers, incident energy "sees" free space as it enters the material and is attenuated as it propagates through a medium which becomes lossier. Instead of having a medium of uniform properties, dielectric constant can be adjusted to be very low at the front surface and increase to a relatively high number at the rear. Consequently, the front-face reflection and the overall thickness, which are required to make the back-face reflection negligible, are greatly reduced, since very high impedance can be tolerated in the lossy material at the rear. Extremely low reflection levels can be achieved in this manner, resulting in performance more than 60 dB below the incident energy levels under best conditions.

An example of broadband absorber is graded-dielectric absorber. It consists of different layers of conductive carbon which is pyramidal in shape in polyurethane foam. The principle of operation of such absorber is quite different from that of the resonant type. Absorption is achieved by a gradual tapering of impedance from that of free space to a highly absorbing state (lossy state). If the transition from free space to a highly absorbing state is done smoothly, little reflection from the front face will result, making it a good absorber. Absorption levels greater than 50 dB can be obtained with pyramids of many materials with different thicknesses (thickness in terms of wavelength) with the graded-dielectric absorber. Good levels of reflectivity reduction (greater than 20 dB) can be achieved in the materials of thickness less than 1/3 wavelength. In this case, very open-celled (10 pores per inch) foam is used. In this type of absorber, gradual transition, i.e., free space to a highly absorbing state, is achieved via a conductive carbon coating. These types of absorbing materials are used for "broadband" absorption characterized by a gradual tapering from the impedance of free space to that of the medium over/above appreciable thickness.

2.4.3.5.3 Magnetic Absorber

Magnetic absorbers alter the permeability for the absorption of microwave waves, e.g., ferrites, iron and cobalt-nickel alloys. For analysis of microwave absorption, the magnetic properties such as permeability (which has a complex number with real and imaginary parts) are used. The strong magnetism properties of magnetic absorbers make them a good microwave absorber. Magnetic absorbers, particularly ferrites, are special kinds of absorbing materials due to their dielectric and magnetic losses in microwave frequency band (Figure 2.7). Ferrites are complex solids, represented by $M+2OFe_2O_3$; where, M+ is the ion of a divalent metal such as cobalt (Co), nickel (Ni), zinc (Zn), cadmium (Cd), iron (Fe), manganese (Mn), chromium (Cr), copper (Cu), etc., or a mixture of these. The specific resistance of ferrites is very high (of the order of 107–108 ohm m), which is 1014 times as high as those of metals.

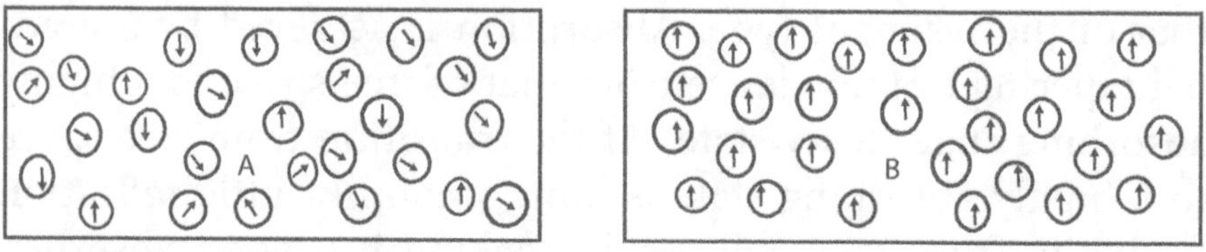

Figure 2.7 Schematic representation of the effect of microwave absorption by magnetic materials (A) before and (B) after absorbing the microwave.

The relative permittivity for ferrites is of the order of 10–15 and loss tangent **tan δ = 10⁻⁴** (low loss at microwave frequencies). The relative permeability of ferrites is several thousand. As the ferrites are transparent to the electromagnetic waves, an electromagnetic ray propagating through the ferrites encounters strong interaction with spinning electrons and gives rise to desirable magnetic properties in ferrites. Spherical ferrite and carbonyl iron powders are among the most popular conventional magnetic fillers. Usually, the microwave-absorbing properties of the ferrites are dominated by the magnetic and dielectric losses. The material that exhibits both dielectric and magnetic parameters is known as four-parameter material, which plays an important role as an absorber.

2.4.3.5.4 Dielectric Absorber

Dielectric absorber is material in which the electrical and magnetic properties are altered to allow absorption of microwave energy at discrete or broadband frequencies. Commonly used dielectric materials as absorbers are foams, plastics and elastomers. These materials do not have magnetic properties, giving them permeability of '1'. High dielectric materials, such as carbon, graphite and metal flakes, are used to modify the dielectric properties of such materials so that they are able to absorb better.

For dielectric absorbers, loss is primarily generated via the finite conductivity of the material. Incident EM waves impinging upon a conductive surface induces current as the electric field of electromagnetic wave interacts with mobile electrons within the material. Additional loss can also occur via molecular polarization phenomena, such as dipole rotation.

2.4.3.5.5 Metal Absorber

In most of the absorbers, the metal particles or the conducting material particles are used as filler. These particles alter electrical properties such as permittivity and conductivity of the material.

In practice no conductor is perfect, i.e., it has a finite conductivity σ. This imperfectness of the conductor results in a very small portion of energy entering the conductor from free space (air), i.e., the second medium. However, due to the high value of σ, the waves are highly attenuated in conductor.

Propagation constant for a conductor is

$$k = \eta \ \sigma m$$

where, **η** is the intrinsic impedance and **σm** is conductivity of medium.

$$\text{As, } \boldsymbol{\eta = (\mu/\varepsilon)\ 1/2}$$

We have,

$$\alpha = \sigma m\ /2(\ \mu/\varepsilon\ )1/2 = \sigma m\ /2.\ \eta$$

Since, **α** is a function of **σ**; therefore, the attenuation depends on the nature of the metal How and Vittoria [20]. The property of the metal to attenuate the fields is expressed in terms of quantity defined as the depth of the penetration or skin depth. It is defined as the distance within the metal in which the amplitude of the traveling wave is attenuated to 1/e or 36.8% of its value just above the metal surface. Then by definition, if $\boldsymbol{d}$ is the skin depth, then

$$\alpha d = 1 \text{ or } d = 1/\alpha$$

For copper and silver, $\boldsymbol{d}$ values are 0.00667 mm and 0.00637 mm respectively at 100 MHz frequency. Thus, in good conductors, such as copper, silver, etc., the wave is attenuated to a negligibly small value within a few thousands of a centimeter. In addition to the above-discussed absorbers, the composite materials also play an important role as an absorber. These materials allow convenient use on surfaces and good control over mechanical properties. These properties along with the variation of microwave properties, depend on the proper selection of matrix material and different inclusions, which are either dielectric, conductive, or ferromagnetic [15–18]. These magnetic absorption materials made by dispersing magnetic fillers in an insulating matrix, continue to play a leading role in the application of microwave absorption [19, 20]. Although many kinds of ferrites with particle sizes from micron to millimeter magnitude have been utilized for this purpose but the coatings of these absorbers are still too thick and heavy to meet practical demands [21].

2.4.3.5.6 Active Cancellation

Active cancellation is a modern method of radar jamming system that involves the sampling of an incoming radar signal, analyzing it, then returning this signal slightly out of phase, thus "canceling" it out.

2.5 Radar Cross Section (RCS)

Radar cross section is the measure of a target's ability to reflect radar signals in the direction of the radar receiver, i.e., per unit solid angle. Not all of the radiating energy falls on the target. RCS is not directly proportional to detection range because of radial scattering effects and canonical beam of radar. The RCS is seen as the percentage of power reradiated (scattered) by the target and the ratio of power scattered back in radar direction. Reducing the RCS alone is not enough. There is a need to minimize the RCS that can be achieved by addressing the design of aircraft.

2.6 Shaping of Stealth or Improved Aerodynamics by Reducing RCS

Another way to reduce the RCS is by propulsion subsystem shaping, i.e., creating fluidic spout for thrust vectoring in aircraft jet engines. Such spouts produce a much lower RCS (radar cross section). Aircraft shape makes a significant difference in detectability. A stealth aircraft requires a flat or rectilinear shape, which at the same time is oblique to the incoming waves, compact and has smoothly blended external geometry to achieve a continuously varying curvature.

- Conventional aircraft use a rounded nose cone that supports a basic principle of aerodynamics.
- The stealth aircraft is made up of flat surface and very sharp edges.
- Radar signals heating the stealth plane are scattered in all directions by this.
- Metallic surfaces generally reflect the radar signals.
- Therefore, stealth aircraft are coated with radar absorbent materials (RAM), which, as mentioned above, deflect and absorb incoming radar waves and reduce the detection range.
- Radar absorbent layer is present below the surface coating of aircraft using corner reflector.

So far, the surface modification of aircraft is discussed; but internal construction also needs attention because radar waves penetrating the surface of the aircraft get trapped in special internal structures known as re-entrant triangles and bounce off the internal faces, losing energy. Such structures are buried in the engines within the wing or fuselage of stealth aircraft. Planform alignment is also often used in stealth designs, which returns a radar signal in a direction away from the radar emitter.

2.7 Reducing the IR Signature

Every atom of matter continuously sends out electromagnetic radiation at an infrared (IR) wavelength which corresponds to its temperature. The engine exhaust has carbon dioxide that produces most of the IR signature at 4.2 microns. The exhaust from stealth aircraft dissipates fairly quickly after leaving the aircraft. However, this needs to be reduced.

Exhaust plume from stealth aircraft creates significant infrared signature, which needs to be reduced. For this purpose, the interest is near IR region, i.e., shorter than 10 μm. The major IR emission sources from an aircraft are tailpipe region, turbojets having EGTs of 1000 °C, exhaust gas plume, hot part of engine, aircraft skin or surface that creates frictional heating, reflected and reemitted sunlight. Techniques used for reducing IR signature are:

- Having a noncircular tail pipe that minimizes the exhaust cross-sectional volume and maximizes the mixing of the hot exhaust with cool ambient air.
- Injecting cool air into the exhaust flow to boost this process.
- Venting the jet exhaust above the wing surface to shield it from observers below.
- Cooling the exhaust gas to the temperatures where the brightest wavelengths it radiates on are absorbed by atmospheric carbon dioxide and water

vapor, thus reducing the IR visibility of the exhaust plume.

- Circulating coolant fluid to reduce the exhaust temperature, such as fuel inside the exhaust pipe, where the fuel tanks serve as heat sinks cooled by the flow of air along the wings. For this, engines can be fitted with flow mixers to blend the cold bypass air with the hot air that passes through the combustor and the turbine.
- Adapting exhaust geometry to a wide and flat shape.
- Engineering the interaction between the exhaust stream and the airflow over the aircraft to create an additional vortex which further promotes mixing.

2.8 Muffling Aircraft Noise

Initially, slow-turning propellers were used to avoid being heard by enemy troops below. Though supersonic and jet-powered stealth aircraft have high acoustic signature, they do not need much consideration as their speed is very high and they fly at extremely high altitude. The rotor blades of stealth helicopters create much noise, which can be reduced by modulating the blade spacing.

In addition to reducing IR and acoustic emissions, other radiating detectable energy from onboard radars, communication systems, or radio frequency (RF) leakage from electronics enclosures are also avoided.

2.9 What Is Counter Stealth and Why Is It Used?

When the capabilities of stealth aircraft came into existence, the need to develop anti-stealth technologies was felt. Today the potential counter-stealth approaches include passive/multistatic radars, very low frequency radars, over-the-horizon radars and sensitive IR sensor systems, etc.

The radars work in three different bands, i.e., X-band radar, S-band radar or L-land radar. Initially, stealth was designed to counter X-band radars. But now there are radars that work in S-band or L-band and therefore X-band radars are useless; hence, there was a need to develop the stealth of different wavelengths of the radar. Radar working in L-band produces wavelengths with size relative to the aircraft itself and should exhibit scattering in the resonance region rather than the optical region, otherwise the existing stealth will change from sightless to visible.

2.9.1 LIDAR (Light Detection and Ranging)

The attributes of LIDAR are its short wavelength, high beam quality, high directionality and high measuring accuracy. In addition, LIDAR also holds higher resolution and counter-jamming ability due to its coherence property and ultimately high frequency. It is a multi-band and multi-static anti-stealth technology. All these properties help in identifying targets, posture displaying and orbit recording.

2.9.2 Multi-Band 3D Radar

Multi-band 3D radar is the latest counter-stealth radar which was developed by the Russian Defense Ministry in 2008. As the name suggests, it is a pack of three to four discrete radars which is controlled by a single processing and command unit. It can search and track functions of low RCS targets with the X-band; and provides nice tracking using L-band. It is difficult to jam counter-stealth radar because it has a passive angle tracking capability against jammers. Passive anti-stealth radar does not use reflected energy and, hence, is more accurately denoted as electronic support measure (ESM) system. It can detect radiation signals emitted by different sources, such as radio broadcasts or mobile phone networks, and helps in finding the location of an object. Such radar that can find stealth aircraft has been developed by European Aeronautic Defense and Space Company (EADS). It is difficult to trace this type of radar because passive radar stations don't radiate anything.

2.9.3 Quantum Radar

A patent has been obtained by defense contractor Lockheed Martin regarding quantum radar, which is based on a remote-sensing method based on quantum entanglement. This system is expected to provide better resolution and higher detail than classical radar. It is expected to use photon entanglement to allow several entangled photons to function like a shorter wavelength to detect small details while having an overall longer group wavelength that allows long-distance transmission.

2.10 Nanotechnology for Improved Stealth Performance

Stealth means the act of trying to hide or evade detection of the aircraft. However, the present-day broad concept of stealth technology incorporates and refers to a series of technologies and design features beyond the conventional radar concept. Here we look at the modern vision of aviation design requirements like faster, miniaturized, highly maneuverable, self-healing, intelligence-guided, smart, eco-friendly, and lightweight designs warranted by extraordinary mechanical and multifunctional materials, along with some support systems for stealth technology.

Nanotechnology has found potential use of nanomaterials in enhancing low observability along with light weight, high strength and toughness. Many nanoparticle coatings are capable of making corrosion-resistant aircraft, that gives durability and reduces maintenance. Such aircraft have increased payload and are also cheaper and safer. Moreover, nanotechnology is also applied for improved electronics, displays with low power consumption, sensors, filters for air purification, nanomaterials in tires and brakes, etc.

Nanomaterials that have found maximum application in developing stealth include carbon nanotubes (CNT) and their nanocomposites, ferrites and their nanocomposites, ceramic compounds and composites and silicon conjugated with other nanoparticles.

The direct role of nanotechnology in advancing stealth has been in developing microwave absorbing nanomaterials or RAM (radar absorbing material). Hence, let us first discuss microwave absorption using nanomaterials.

2.10.1 Nanomaterials as RAM or Microwave Absorber

As discussed above, RAM should be able to increase and decrease the reflection and the bandwidth of the material should have high absorption capacity. Carbon, especially CNT conjugated with different materials, is showing a lot of promise. A number of microwave absorbing or radar absorbing nanomaterials, such as ferrites [22–24], ferromagnetic composites [25], conducting fibers [26], and carbon nanotubes [27], have been explored for microwave absorbing coatings on the surface of military stealth aircraft to avoid detection by enemy radar. Microwave absorbing nanomaterials are usually prepared as a suspension of colloidal nanoparticles in a solvent, in which electrostatic or steric repulsive forces prevent aggregation of nanoparticles [28]. There is a need to develop a nanofilm deposition method for stealth.

Presently, the most common microwave absorbers that are in use are a number of ferromagnetic materials. However, the technical requirements for the absorbing material, such as wide frequency range, zero external magnetic field, and thin absorption layer, limits its use in the required microwave frequency range of a few gigahertz.

Nanocrystalline materials possess very high surface area; moreover, the nanocrystalline structure may form multiple magnetic domains and the coercive force of the material will greatly increase, which may lead to large hysteresis attenuation and thus the absorbing properties can be greatly improved. Therefore, nanostructured materials are being considered for absorbing microwaves in the GHz frequency range. In addition, many other unique chemical and physical properties of nanoparticles, such as their volume-to-weight ratio, lightweight and strong absorption, offer their applicability as microwave absorber.

2.10.1.1 Nanoferrite Absorber

Looking at the advantages of nano-sized materials, nano-sized ferromagnetic particles were considered by many to extend the applicability of ferrites already in use and allow further fine-tuning of the absorbing characteristics. No wonder that ferrite materials are one of the most widely used magnetic materials for application in wireless communication and wireless sensing systems. Some ferrite materials with strong magnetocrystalline anisotropy fields can extend these applications to tens of GHz range while reducing the size, weight and cost. The ferrite materials with strong magnetocrystalline anisotropy field are metal/non-metal-substituted iron oxides oriented in low crystal symmetry. The ferrite materials studied include M-type hexagonal ferrites such as barium ferrite ($BaFe_{12}O_{19}$), strontium ferrite ($SrFe_{12}O_{19}$), epsilon phase iron oxide (ε-Fe_2O_3), substituted epsilon phase iron oxide (ε-$Ga_xFe_{2-x}O_3$, ε-$Al_xFe_{2-x}O_3$) and many more.

Since the evaluation of absorbing characteristics depends on the desired properties, emphasis can be given to various parameters such as thickness, weight, wide band absorption in some frequency ranges, etc. Many challenges were faced when using nanoferrites, e.g., for the development of an electromagnetic wave absorber for the desired matching frequency, because the ferrite materials used are manufactured by a complex process involving conditions like controlled sintered temperature, pressure, and a specific ratio of composite materials. These challenges make it difficult to develop a material which can absorb the microwave in a wide range of the frequency band. Also, these magnetic absorbers tend to be heavy; their virtue lies in their ability to provide extended frequency performance with reasonable material thickness of 1–5 mm thickness [3]. Unfortunately, these materials need a thick coating to meet practical demands. Therefore, it was felt that there was a need to develop materials so that even thin coating should be sufficient to absorb microwaves [29].

Various compositions of nanocrystalline ferrites have been synthesized and tried for this purpose.

2.10.1.1.1 $Li_{0.5}Mn_{x//2}Fe_2O_3$ Nanocomposite

Anwar and Maqsood [30] prepared a soft polycrystalline nanoferrite composed of $Li_{0.5}Mn_{x//2}Fe_2O_3$ by sol-gel auto-combustion method and studied its absorption properties. Here $x = 0.0$, 0.3, 0.6, 0.9, and 1.2. Addition of Mn was found to increase the magneto-dielectric properties; the sample showed low magnetic loss in very high frequency and ultrahigh frequency (VHF and UHF), which makes it useful in miniaturizing the antennas in the frequency range 1 MHz to 1000 MHz. Moreover, it exhibited high reflection loss (RL) of 45 dB for $x = 0$ and relatively wide bandgap. Thus, it has been found to be suitable for the lower microwave (MW) region.

2.10.1.1.2 Barium Hexaferrite/Polyaniline Core-Shell Nanocomposites

This new composite was fabricated by Liu *et al.* [31] by *in-situ* polymerization. They tuned the electromagnetic properties of this nanocomposite by making the desired shell thickness. The proper core/shell thickness made contributions to refining impedance matching. The optimized core-shell nanocomposites with a shell layer of 30–40 have exhibited a broad −10 dB absorption bandwidth of 3.8 GHz. Both the microwave absorbing efficiency and frequency range changed as the shell thickness increased.

2.10.1.1.3 Nanostructured Nickel Ferritev

Jacob *et al.* [23] synthesized nano-sized nickel ferrite using a simple chemical precipitation technique. They showed that the microwave dielectric properties, such as dielectric constant, dielectric loss and microwave heating coefficient, varied with variation of average grain size.

2.10.1.1.4 Layer-by-Layer Self-Assembled Ferrite Multilayer Nanofilms

Layer-by-layer consists of multilayer thin film at the molecular level [32, 33]. For self-assembly of multilayer film, sequential adsorption of oppositely charged polymers is done to get desired thickness, composition, and density of film [34, 35]. Ferrite nanoparticles that have been incorporated into layer-by-layer deposited multilayer thin film through complementary interactions (e.g., electrostatic interactions, covalent bonding, hydrogen bonding, and hydrophobic interactions) include cobalt ferrites [36], zinc ferrites [37], and other mixed ferrites [38].

Heo *et al.* [39] fabricated multilayer thin films of $Co_{0.5}Zn_{0.5}Fe_2O_4$ nanoparticles. The ferrite nanoparticles incorporated were oleic-acid- (OA-) stabilized or ligand-exchanged using bPEI;

the OA-stabilized nanoparticles had a diameter of about 18.5 nm, compared to 7 nm after ligand exchange. The ferrite nanoparticles were deposited onto a Si wafer using an electrostatic-interaction-based layer-by-layer self-assembly method. They claimed that these films have microwave absorbing properties for possible radar absorbing and stealth applications.

2.10.1.1.5 Nano Spinel Ferrite

Spinel ferrite or iron oxide nanoparticles have a heavy mass problem. To solve this problem, they have been conjugated with many inorganic nanoparticles such as Co, Zn, Ni, Mn and also organic polymers. Huang *et al.* [40] prepared cobalt zinc spinel ferrite nanofibers and found this lightweight composite to be efficient microwave absorber. The synthesis of $CoxZn_{(1-x)}Fe_2O_4$ (x = 0.2, 0.4, 0.6, 0.8) ferrite nanofibers was done by electrospinning method. Their findings included enhanced saturation magnetization and coercivity; that was achieved by tuning the Co^{2+} content. Moreover, the electromagnetic loss analysis indicated that the $Co_{0.6}Zn_{0.4}Fe_2O_4$ ferrite nanofibers had the strongest microwave attenuation ability. A coating of 15% $Co_{0.6}Zn_{0.4}Fe_2O_4$ ferrite nanofibers showed less than −10 dB reflection loss in the whole X-band and 80% of the Ku-band frequencies. Meanwhile, the surface density was only 2.4 Kg/m^2.

2.10.1.2 Nano-Carbon and Carbon Nanotube (CNT) Composites as Absorbers

CNT/ferrite composites have received maximum attention for use as microwave absorbing materials. Morphologically, CNTs appear as two-dimensional graphene sheet rolled up into a tubular structure. With only one wall in the cylinder, the structure is called a single-walled carbon nanotube (SWCNT) and with many concentric walls with a constant interlayer separation of 0.34 Å is called a multiwalled carbon nanotube (MWCNT).

The unique property of CNT's structure is that it is characterized by a chiral vector (n,m). When m-n/3 is an integer, the resulting structure is metallic; otherwise, it is a semiconducting CNT. This very unique electronic property has led to numerous applications in nanoelectronics.

Various nanoforms of carbon have been synthesized. It is an extremely versatile molecule; it can form linear or zigzag chains, rings, sheets or blocks because it forms strong bond with its own atom (bond energy of C-C bond is 348 kJ/mol). This property leads to formation of a long network of carbon structures. However, sp^3 and sp^2 bonding carbon atoms are the main building blocks of all the carbonaceous structures available so far. Nanocarbons are synthesized from hydrocarbon [41–43] as precursor, which is catalytically decomposed to liberate carbon atoms and constitute carbon materials of different sizes and shapes through optimizing the synthesis parameters (temperature, pressure, reaction duration, catalyst, etc.). These various carbon structures have made a great impact as a radiation absorber [44–47]. SWCNT, MWCNT, CNB (carbon nanobead) and CNF (carbon nanofiber) are all nanoforms of carbon that have shown microwave absorbing capacity when conjugated with different polymers.

2.10.1.2.1 Carbon Nanobeads (CNB)

Carbon nanobead film, synthesized from *Brassica nigra* oil, can absorb microwave in the range of ~75–90%. The Absorption band widths of 0.8 GHz and 0.6 GHz in the frequency range of 15.2–16 and17.4–18 GHz suggested its capability as an absorbing material in the Ku-band of an electromagnetic wave.

To use nanoforms of carbon as microwave absorber it has to be used as composite with any polymer that can facilitate the coating process. To study the microwave absorption, the powder of CNB, CNF or CNT is mixed with acrylic paste. The microwave absorption was not much affected by adding acrylic

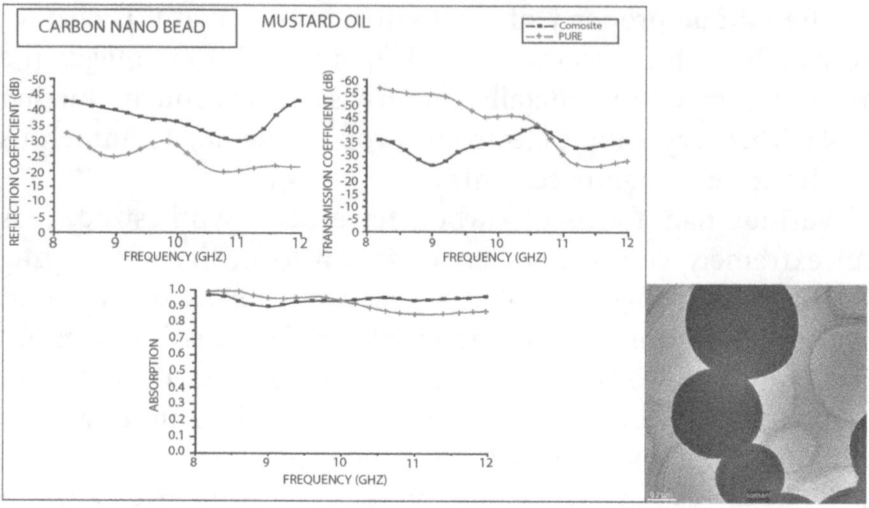

Figure 2.8 (left) Variation of reflection coefficient, transmission coefficient and absorption with mustard oil-based CNB filler in acrylic as a function of frequency in the X-band. (Absorption observed by only CNB is shown in red and with acrylic is shown in black.) (right) TEM image of CNB synthesized from *Brassica nigra* oil.

paste; rather it was more or less the same with or without the acrylic for all three forms of nanocarbon (Figure 2.8).

CNBs show a decrease in absorption at lower frequencies, i.e., up to 11.6 GHz (Figure 2.8).

2.10.1.2.2 Carbon Nanofibers (CNF)

Carbon nanofibers synthesized from linseed oil show absorption above 8.8 GHz, with an appreciable increase in RC corresponding to the low reflection of microwave in the X-band (Figure 2.9) The low reflection by the CNF composites corresponds to the large transmission of up to 11.6 GHz; above this frequency, low transmission is observed with an increase in transmission coefficient (TC).

Carbon nanofiber has a longer length and diameter compared to CNB and MWCNT; hence, the microwave energy loss

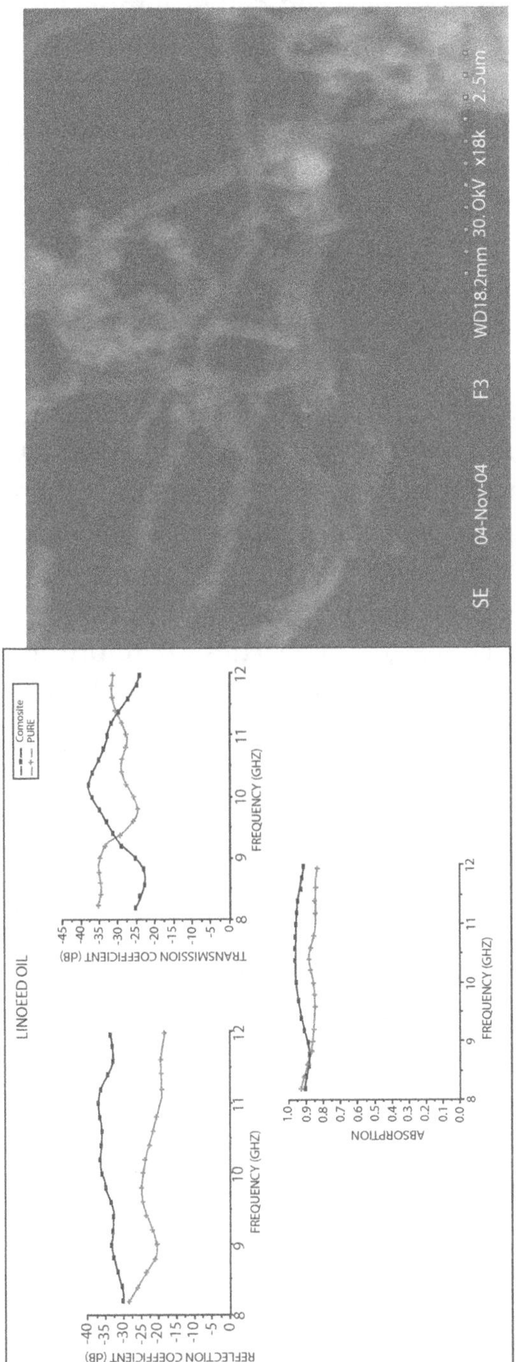

Figure 2.9 (left) Variation of reflection coefficient, transmission coefficient and absorption with linseed oil-based CNF filler in acrylic as a function of frequency in the X-band. (Absorption observed by only CNF is shown in red and with acrylic is shown in black); (right) SEM Image of CNF synthesized from Linseed oil.

inside the CNF is more. This is indicative of more loss of microwaves in CNF composite, where displacement current is more effective than the conduction current. Variation in the absorption with frequency suggests the frequency-dependent behavior of the material. Variation of reflection coefficient, transmission coefficient and absorption by CNF-acrylic composite is a function of frequency in the X-band. (Absorption observed by only CNF is shown in red and with acrylic is shown in black in Figure 2.9).

2.10.1.2.3 Multiwall Carbon Nanotubes (MWCNT)

Apart from having unique electronic property, MWCNTs also exhibit other extraordinary properties such as (i) mechanical properties having Young's modulus over 1 TPa and a tensile strength of about 200 GPa; (ii) thermal conductivity as high as 3000 W/mK; (iii) ideal aspect ratio; (iv) small tip radius of curvature; (v) good emission properties, therefore are excellent candidates for field emission; (vi) CNTs can be chemically functionalized by a variety of atomic and molecular groups to the ends of sidewalls. Polymer-based composites of CNTs exhibit a wide range of Young's modulus, high specific strength, crash resistance and thermal performance.

MWCNT and their composites have been explored for their use as microwave absorbing materials. Biogenically synthesized MWCNT from karanjia oil as precursor and their composite with acrylic has exhibited variation in reflection coefficient, transmission coefficient and **microwave** absorption as a function of frequency in the X-band. (Figure 2.10).

2.10.2 Nanomaterials in Airframe Structure

Nanomaterials are being used and researched for use in three areas, i.e., (i) Airframe Structure, (ii) Aero-Engine Parts and (iii) Aircraft Electro-Communication System. The main

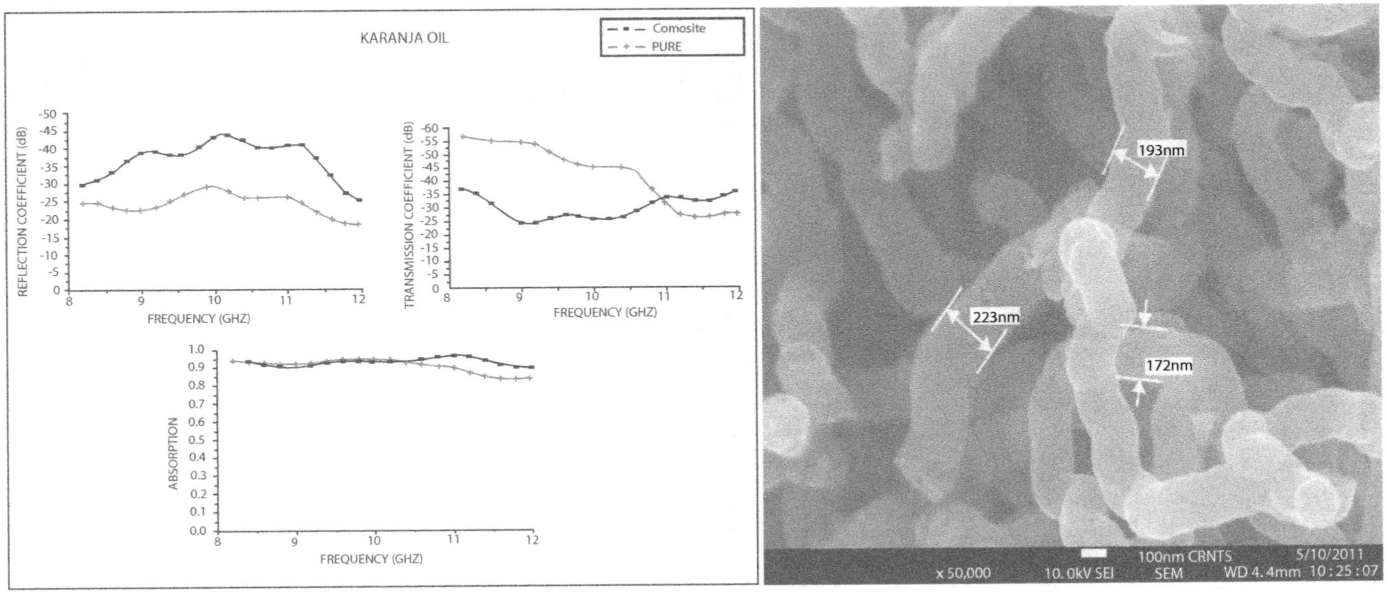

Figure 2.10 (left) Variation of reflection coefficient, transmission coefficient and **microwave** absorption with MWCNT (synthesized from karanjia oil) filler in acrylic as a function of frequency in the X-band. (Microwave absorption by MWCNT is shown in red and by MWCNT + acrylic is shown in black.) (right) SEM Image of CNF synthesized from karanjia oil.

nanomaterial properties required for the design of an airframe structure are that the material be very lightweight and have high strength and toughness. Moreover, the material has to be resistant to corrosion, durable, easy to maintain, easily repaired and reusable. These properties are needed to make faster, miniature, highly maneuverable, self-healing, intelligence-guided, smart, eco-friendly, and lightweight stealth systems. Various nanomaterials are used as fillers to enhance the required properties of polymers that are used in aircraft construction, especially to bring down the weight and the cost. Nanomaterials which can fulfill the requirements of airframe structure are discussed below.

2.10.2.1 Carbon Nanotubes (CNTs) Conjugated with Polymers

The polymers used in making CNT/polymer conjugate function as the matrix material. CNT-based polymer composites, such as CNT conjugated with epoxy, CNT/polyethylene, CNT/polymide or CNT/polypropylene, exhibit a wide range of Young's modulus, high specific strength, crash resistance, conductivity and thermal performance that can be used for airframe structure. According to the studies done by Sarah E. O'Donnel of MITRE Corporation (Center for Advanced Aviation System Development), CNT nanocomposites affect the design, flight performance and efficiency of aircraft that will reflect on better safety and capacity (Table 2.1). Scientists believe that in the future, stealth manufacturers armed with nanotechnology will design cost-effective airframes for mitigated wake vortex circulation, better defense and intelligence purposes.

The tubular structure of CNT enables ballistic electron and phonon transport [55], giving CNTs much higher current carrying and heat conducting capacity (a SWCNT conducts nearly 2 trillion electrons/sec through the ~3 nm nanotube molecule

Table 2.1 Various CNM and their composites used as microwave absorbing materials.

Composite material	**Size**	**Film thickness**	**Freq. range**	**AB. Freq. GH_z**	**RC loss**	**Reference**
CNB-filled EVA & NR base composite (Vulcan XC-72) CB	29 nm	1,8–3.5 mm	8–12 GH_z	12	-3 dB	Das *et al.* [12]
CNT & Benzene composite	70 nm	0,97 mm	8–18 GH_z	11.41	-22.89 dB	Shen *et al.* [48]
CNF & Micro-coiled Paraffin Wax	CNF 500-600 nm & Wax 4-5μm	—	12.4–18 GH_z	12.4–18	tan§ is 0.409 dB 0.407 dB	Du *et al.* [44]
Nano Si/C/N composite powder (Paraffin wax)	20–30 nm	3 mm	8–18 GH_z	8–18	< -8 dB	Zhao *et al.* [49]

(*Continued*)

Table 2.1 Various CNM and their composites used as microwave absorbing materials. (*Continued*)

Composite material	Size	Film thickness	Freq. range	AB. Freq. GH_z	RC loss	Reference
Carbon nanobeads	2 μm	—	8–12 GH_z	8–12	~ -25 to -35 dB	Sharon *et al.* [29]
Hydrogen Plasma in SWCNTs	0.8–2 nm	1 cm	300 MH_z to 30 GH_z	22,20,2,4,2.55	-29.83 dB -34.32 dB -28.68 dB -32.67 dB	(HiPco SWCNTS)
Polyaniline-CNT composite	20–50	10 wt%	8–12 GH_z	8-2–10, 12.4	tan δ >1 i.e. lossy material	Makiff *et al.* [50]
MWCNT/polymer & MWCNT/ varnish composite	1–2 μm agglomerated MWCNTs	CNT/PET–2 mm CNT/varnish – 1 mm	2–18 GH_z	7.5, 15.36	-17.61 -24.27 dB	Fan *et al.* [51]
Polymer composite containing dispersed CNTs	CNT ≤100 nm	2 cm	40 MHz–40 GHz	30,15	σ =1 s/cm leads to good absorption dB	Saib *et al.* [52]

(*Continued*)

Table 2.1 Various CNM and their composites used as microwave absorbing materials. (*Continued*)

Composite material	Size	Film thickness	Freq. range	AB. Freq. GH_z	RC loss	Reference
C-coated Ni Nanocapsules	Ni(C) Nanocapsule	2 mm	2–18 GH_z	13	-32 dB	Zhang *et al.* [53]
SWCNT/ polyurethane Composites	Avg. 5.95 nm	16–20 wt%	8.2–12.4 GH_z	12.4, 11.25,19.3, 9.25	-20 dB	Liu *et al.* [31]
E-glass fabric/ epoxy composite laminates containing CNF & MWCNT	-	-	0.5–18 GH_z	10	-20 dB	Kim *et al.* [53]
CNF	100–200 nm	=	8–12 GH_z	8.6–10.20	-27.42 to -39.90 dB	Kshirsagar *et al.* [47]

diameter) than copper (a ~ 3 mm cross section) Cu conducts approximately 2 million electrons/sec) [56].

2.10.2.2 Nanoclay Reinforced Polymer Composites

Montmorillonite is clay, which is a very soft phyllosilicate group of minerals. Clay is in the form of microscopic crystals. Chemically, it is hydrated sodium calcium aluminum magnesium silicate hydroxide $(Na,Ca)_{0.33}(Al,Mg)_2(Si_4O_{10})(OH)_2 \cdot nH_2O$.

Whereas *nanoclays* are derived from montmorillonite, a mineral deposit that has layered structure of dimension around 1 nm thick and a specific surface area of 700–800 m2/g13. Nanoclay is used to improve modulus and tensile strength as well as barrier properties, flame resistance, and thermal properties of many plastics. Nanoclay may be nanoparticles of layered mineral silicates or can be montmorillonite, bentonite, kaolinite, hectorite, and halloysite. Nanoclay is reinforced polymer composites having unique barrier properties and they are thermal and flame retardant. Timmerman *et al.* [57] have shown that laminates prepared from carbon fiber/epoxy composites reinforced with nanoclay show significant inhibition in cryogenic microcracking.

2.10.2.3 Metal Nanoparticle Incorporated Composites

Airframe structure is made mostly by using bulk metals. The nanoparticles of some of these bulk metals are now well known for their novel properties that are not found in their bulk counterparts.

Metal nanoparticles which have found application in airframe structure include:

(i) Nano-silica conjugated with rubber compounds are used to prepare gaskets and sealants which may find applications in aero parts or engines;
(ii) SiC nanoparticles reinforced alumina;

(iii) Nano-chromium-based corrosion inhibitors are being developed for protection of aluminium metals or aero structures;

(iv) Yttria-stabilized nano-zirconium that can facilitate crack healing, improve high-temperature, strength and creep resistance, which are better than monolithic ceramics;

(v) TiN nanocrystallites embedded in amorphous Si3N4, which has found its use in wear-resistant coatings and low friction applications of aircraft;

(vi) Nanopowder of copper, aluminium, and iron are being used for preparing conductive plastics to be used in various components of aircraft where electrostatic discharge of EMI shielding applications is required.

2.10.3 Nano-Metal Coatings for Aero-Engine Parts

One of the major trends in application of nanomaterials in stealth concerns nano-coating metals for better durability and corrosion resistance. The coatings are generally used for protecting the structures and surfaces of the aircraft from harsh environments, i.e., extreme temperatures, extreme climates, corrosion, abrasion and wear of engine parts; preferably a nano-coating that can stand high engine-operating temperatures so as to offer improved performance.

Metal nanoparticles and metal nanoparticles incorporated composites have extraordinary electrostatic discharge and electromagnetic interference (EMI) shielding properties, that can be used for coating the airframe structure or aero-engine parts to make them resistant to lightning strikes.

(i) Nano-metals used as anti-corrosion material include silicon oxide, boron oxides, cobalt and phosphorous nanocrystals.

(ii) Nano-metal coatings used on turbine blades and other mechanical components which have to withstand high temperatures and friction wear include crystalline carbide, diamond-like carbide, metal dichalcogenide, nitrides, metals, and various ceramics. Such tribological coating improves the efficiency of the engines. Moreover, nanocomposite coatings made of CNT, nanographite, nano-aluminum containing polymer coating are used for electrostatic discharge, EMI shielding and low friction applications of aircraft surfaces.

2.10.4 Nanomaterials for Electro-Communication Component of Aircraft

Demands for modern stealth is directing scientists to work in the area of miniaturization of components and devices and the use of quantum effect of the wave-like behavior of the electrons. Requirements envisaged for electro-communication to be used in an aircraft are (i) nanoparticles for high density reliable data storage media, (ii) nanoparticles for supercapacitors and (iii) MEMS (micro-electromechanical systems) and NEMS (nano-electromechanical systems) for fuel management.

2.10.4.1 Nanoparticles for Data Storage Media

Data storage is done by magnetic recording. Prior to the 1980s, an attempt was made to store data using thin-film head and advanced data coding; which was modified during the 1990s when magneto-resistive head (MR) and thin-film recording media were used. By the late 1990s, the storage density of the hard disk took a giant step forward through the introduction of a giant magneto-resistive (GMR) reading sensor [58]. By late 2006, Hitachi Global Storage Technologies achieved a storage density (density per unit area) of 345 Gbits/in2 in perpendicular recording. Nonetheless, the need for accelerating more information capacity still exists for ultrahigh-density data storage media. Superparamagnetic limit (where even ambient thermal energy reverses the recorded magnetization) has been seen as a possible limit and has become a concern for the hard disk design. To overcome this superparamagnetic limit, the development of exchange bias magnetic materials is being considered. Anti-ferromagnetically coupled (AFC) magnetic media layers are needed as exchange bias material. This anti-ferromagnetic material should have high Néel temperature (i.e., temperature at which thermal energy becomes large enough to destroy the macroscopic magnetic ordering within the material). Large magneto-metal oxide-based exchange bias nanomaterials offer great possibilities for future improvements in data storage, transfer, and retrieval; and crystalline anisotropy and good chemical and structural compatibility with ferromagnetic films. NiO and CoO are considered to be such good antiferromagnetic exchange bias material. The next step is the use of nanoparticles for ultrahigh data storage media.

Currently, the use of nanoparticles in magnetic storage media is very limited. Efforts are being made to use nanoparticles to form superlattices so as to reduce the size of a bit cell from multiple grains to single particles that will increases the areal density.

Metal-oxide-based exchange bias nanomaterials have been found to improve the data storage, transfer, and retrieval. Magnetic nanoparticles (iron oxide nanoparticles, i.e., Fe_2O_3 and Fe_3O_4) incorporated polymer films and composites are also being tried for various data storage media.

Another suitable material for high-density data storage media is FePt and CoPt magnetic nanoparticles film [59]. It has high anisotropy, which makes it thermally stable down to a particle size of about 3 nm. A longitudinal recording of 100–200 GB per square inch has been obtained by standardizing the head and media dimension ratio. It is suggested that an areal density of about 1 TB per square inch can be achieved with perpendicular recording; and if self-assembled FePt is used, it can reach 10–50 TB per square inch in single-particle-per-bit patterned recording. Scientists are now focusing on nanotechnology for achieving the hard disk design.

2.10.4.2 Nanoparticles for Supercapacitors

A supercapacitor is an electrical double-layered capacitor (EDLC) with moderate energy and high power density. The term "supercapacitor" means a device utilizing the storage of charge in the electric double layer between an electrolyte and electrochemically inert electrode. To be more precise, an electric double layer capacitor can be defined as a device using induced ions between an electronic conductor such as activated carbon and an ionic conductor like organic or aqueous electrolyte. It is an intermediate system between electrochemical batteries that can store high energy associated with low power and dielectric capacitors, which can deliver high power in a few milliseconds.

(i) Recently developed carbon nanotubes (CNTs) of different nano-sized carbonaceous materials like carbon nanotubes, carbon nanofibers, carbon nanobeads, etc., play an important role in the development of electrodes for energy devices and storage systems because of their superior thermal, electrical, structural and mechanical properties. CNT is thermally stable up to 2800 °C in vacuum, and has thermal conductivity about twice as high as diamond and electric-current-carrying capacity 1000 times higher than copper wires [60]. The mesoporous nature of CNTs determines their electrochemical properties and facilitates the transportation of ions while playing an adsorption rule. Due to the accessible electrode/electrolyte interface and low electrical resistance of nanotubes, high power devices are expected to be built with electrodes from these materials [61]. Capacitance values strongly depend on the type of nanotube. Additional capacitance has been obtained by coating MWCNTs with conducting polymers, e.g., polypyrrole [62].

(ii) Ceramic nanoparticles like barium titanate and barium strontium titanate are also used for making supercapacitors.

2.10.4.3 MEMS and NEMS for Fuel Management

Following the tremendous success that MEMS (micro-electromechanical systems) have offered for fuel management, it has been realized that NEMS (nano-electromechanical systems) offer a greater possibility of developing a standard fuel management unit which controls the fuel control in aero-engines.

Since defense is an area that functions confidentially, many developments in the NEMS arena are not available. There has been significant interest in a direct-write, multi-beam, electron-beam lithography tool having sub-10-nm spot size control. This tool will help in developing more cost-effective, high-performance, low-volume integrated circuits (ICs) that are highly customized, application-specific integrated circuits (ASCIs) for use in Maskless Direct-Write Nanolithography for the Defense applications program. It will provide the technology for low-volume NEMS and nano-photonics to enable incorporation of state-of-the-art semiconductor devices in new military systems.

Because CNTs are known to have ballistic electron transport and a huge current-carrying capacity, they are of great interest for future nanoelectronics, especially because they provide desirable electrical properties for polymer matrix composites.

The other advantages of using CNT is its application in electronic components, logic and memory chips, sensors (both physical and chemical), catalyst support, adsorption media, actuators, etc.

2.10.4.4 Other Applications of Nanotechnology in Supporting Advanced Stealth Systems

Some other possible applications for nanotechnology in the aerospace industry are:

- ***Nanomaterials for Energy Storage Devices*** – CNTs have been proved to be potential material for hydrogen storage. The extensive work of Sharon's group has shown that CNT can not only adsorb hydrogen but can also release it in a controlled way to generate power.
- ***Nanomaterials for Ballistic Armor*** – Ballistic armor is used for protection of personnel as well as aircraft from various projectiles. CNTs with their high ballistic resistance capacity, high energy absorption capacity and multi-hit resistance are accepted as being suitable material for making ballistic armor.
- There are many more possible applications of nanotechnology which are still to be explored.

2.11 Conclusions

Though stealth and anti-stealth technology is an interesting topic, most of the research and technology development done in this area is secretive and proprietary in nature, and therefore it is difficult to know exactly what sort of new advances lie in the future. To date, no advancement has been made against quantum radar because of its technology and LIDAR because of its higher accuracy and weather independence. Though all the methods of reducing RCS have been cracked and stealth technology has gone as far as it can, the fact remains that stealth aircraft could be downed, so they are all vulnerable and there is a need to develop counter-stealth technologies. One of the futuristic options seems to be the use of nanomaterials and nanotechnology. This chapter has touched upon this aspect.

References

1. Emerson, W.H., *IEEE Trans. Antennas Propag.*, 21, 4, 484–490, 1973.
2. Ramasamy, D., *Proceedings of INCEMIC 97, 7B-7, IEEE Conference, New Jersey*, pp. 459–466, 1997.
3. Pinho, M., Silveira, P., Gregori, M.L., *Eur. Polym. J.*, 38, 2321–2327, 2002.
4. Meshram, M.R., Agrawal, N.K., Sinha, A.B., *Microw. Opt. Technol. Lett.*, 36, 5, 252–255, 2003.
5. John, D. and Washington, M., *Aviat. Week Space Technol.*, 129, 28–29, 1998.
6. Howarth, D., *Discourse*, 166 pages, Open University Press, Buckingham UK, 2000.
7. Chung, D.D.L., *Carbon*, 39, 279–285, 2001.
8. Barna, E., Bommer, B., Kursteiner, J., Vital, A., Trzebiatowski, O., Koch, W., Schmid, B., Graule, T., *Compos. Part A*, 36, 473–480, 2005.
9. Evans, D., Zuber, K., Murphy, P., *Surf. Coat.*, 206, 3733–3738, 2012.
10. Shi, X. and Croll, S.G., *J. Coat. Technol. Res.*, 7, 73–84, 2010.
11. Pozar, D., *Solutions Manual for Microwave Engineering*, 4th edition, John Wiley & Sons, Inc. New York, Chichester, Weinheim, Brisbane, Singapore, Toronto, 2011.
12. Das, A. and Das, S.K., *Microwave Engineering*, Tata McGraw Hill, New Delhi, 2000.

13. Stonier, R.A., Stealth aircraft and technology from World War II - the Gulf. *SAMPE J*, 27, 4, 9–17, 1991.
14. Bigg, D.M., *Polym. Eng. Sci.*, 17, 842, 1997.
15. A. Korn, Means for altering the reflection of radar waves. U.S. Patent 2436578, Feb. 24, 1948.
16. B.C. Pratt, Electromagnetic radiation - Absorptive article. U.S. Patent 2996710, Aug.15, 1961.
17. O. Halpern, Method of reducing reflection of incident electromagnetic waves. U.S. Patent 3007160, Oct. 31, 1961.
18. Vladimir, B.B., *IEEE Trans. Magn.*, 40, 3, 1679–1684, 2004.
19. Yang, Y., Zhang, B., Xu, W., Shi, Y., Jiang, Z., Zhou, N., Gu, B., Lu, H., *J. Magn. Magn. Mater.*, 256, 129–132, 2003.
20. How, H. and Vittoria, C., *J. Appl. Phys.*, 69, 5183, 1991.
21. Amin, M.B. and James J, R., *Radio Electron. Eng.*, 51, 209–218, 1981.
22. Jeong, G.M., Choi, J., Kim, S.S., *IEEE Trans. Magn.*, 36, 5, 3405–3407, 2000.
23. Jacob, J., Khadar, M.A., Lonappan, A., Mathew, K.T., *Bull. Mater. Sci.*, 31, 6, 847–851, 2008.
24. Zhao, D.L., Lv, Q., Shen, Z.M., *J. Alloys Compd.*, 480, 2, 634–638, 2009.
25. Nie, Y., He, H., Zhao, Z., Gong, R., Yu, H., *J. Magn. Magn. Mater.*, 306, 1, 125–129, 2006.
26. Li, W., Qiao, X., Zhao, H., Wang, S., Ren, Q., *J. Nanosci. Nanotechnol.*, 13, 2, 793–798, 2013.
27. Wadhawan, A., Garrett, D., Pérez, J.M., *Appl. Phys. Lett.*, 83, 13, 2683–2685, 2003.
28. Eberbeck, D., Wiekhorst, F., Steinhoff, U., Trahms, L., *J. Phys. Condens. Matter*, 18, 38, S2829–S2846, 2006.
29. Sharon, M., Pradhan, D., Zacharia, R., Puri, V., *J. Nanosci. Nanotechnol.*, 5, 12, 2117–2120, 2005.
30. Anwar, H. and Maqsood, A., *Electron. Mater. Lett.*, 9, 5, 641–647, 2013, doi: 10.1007/s13391-013-2239-7.
31. Liu, Z., Bai, G., Huang, Y., Ma, Y., Du, F., Li, F., Guo, T., Chen, Y., *Carbon*, 45, 821–827, 2007.
32. Decher, G., *Science*, 277, 5330, 1232–1237, 1997.

33. Shiratori, S.S. and Rubner, M.F., *Macromolecules*, 33, 11, 4213–4219, 2000.
34. Wood, K.C., Boedicker, J.Q., Lynn, D.M., Hammond, P.T., *Langmuir*, 21, 4, 1603–1609, 2005.
35. Hong, J., Kim, B.S., Char, K., Hammond, P.T., *Biomacromolecules*, 12, 8, 2975–2981, 2011.
36. Ibrahim, A.M., El-Latif, M.M.A., Mahmoud, M.M., *J. Alloys Compd.*, 506, 1, 201–204, 2010.
37. Yusoff, A.N. and Abdullah, M.H., *J. Magn. Magn. Mater.*, 269, 2, 271–280, 2004.
38. Tyagi, S., Baskey, H.B., Agarwala, R.C., Agarwala, V., Shami, T.C., *Ceram. Int.*, 37, 7, 2631–2641, 2011.
39. Heo, J., Choi, D., Hong, J., Layer-by-layer self-assembled ferrite multilayer nanofilms for microwave absorption. *J. Nanomater.*, 8 pp. Article ID 164619, 2015, http://dx.doi.org/10.1155/2015/164619.
40. Huang, X., Zhang, J., Xiao, S., Chen, G., *J. Am. Ceram. Soc.*, 97, 5, 1363–1366, 2014.
41. Tkachev, A.G., *Chem. Petrol. Eng.*, 43, 5–6, 351–354, 2007.
42. Rud, A.D., Perekos, A.E., Ogenko, V.M., Shpak, A.P., Uvarov, V.N., Chuistov, K.V., *J. Non Cryst. Solids*, 353, 3650–3654, 2007.
43. Sharon, M., Sharon, M., Kshirsagar, D.E., *Carbon Nano Forms and Applications*, 1st edition, Monad Nanotech. Pvt. Ltd., Mumbai, 2007.
44. Du, J., Sun, C., Bai, S., Ge, S., Ying, Z., Cheng, H.M., *J. Mater. Res.*, 17, 5, 1232–1236, 2002.
45. Kwon, S.K., Ahn, J.M., Kim, G.H., Chun, C.H., Hwang, J.S., Lee, J.H., *Polym. Eng. Sci.*, 42, 11, 2165–2171, 2002.
46. Kshirsagar, D.E., Puri, V., Sharon, M., Sharon, M., *Carbon Sci.*, 7, 4, 245–248, 2006.
47. Kshirsagar, D.E., Puri, V., Sharon, M., Sharon, M., *Synth. React. Inorg. M.*, 37, 477–479, 2007.
48. Shen, J., Huang, W., Wu, L., Hu, Y., Ye, M., *Compos. Sci. Technol.*, 67, 3041–50, 2007.
49. Zhao Dong-Lin, Z. and Znou, W., *Physica E Low Dimens. Syst. Nanostruct.*, 9, 4, 679–685, 2001.

50. Makeiff, D.A. and Huber, T., *Synth. Met.*, 156, 497–505, 2006.
51. Fan, Z., Luo, G., Zhang, Z., Zhou, L., Wei, F., *Mater. Sci. Eng. B*, 132, 85–89, 2006.
52. Saib, A., Bednarz, L., Daussin, R., Bailly, C., Xudong, L., Thomassin, J.M., Pagnoulle, C., Detrembleur, C., Jérôme, R., Huynen, I., *IEEE Trans. Microw. Theory Tech.*, 54, 6, 2745–2754, 2006.
53. Zhang, X.F., Dong, X.L., Huang, H., Liu, Y.Y., Wang, W.N., Zhu, X.G., Lv, B., Lei, J.P., *Appl. Phys. Lett.*, 89, 053115, 2006. https// doi.og/10.1063/1.2230065.
54. Kim, M.-G., Hong, J.-S., Kang, S.-G., Kim, C.-G., *Compos. Part A-Appl. S.*, 39, 647, 2008.
55. Baughman, R.H., Zakhidov, A.A., De Heer, W.A., *Science*, 297, 787–792, 2002.
56. Yao, Z., Kane, C.L., Dekker, C., *Phys. Rev. Lett.*, 84, 2941–2944, 2000.
57. Timmerman, J.F., Brian, S., Hayes, J., Seferis, C., *Compos. Sci. Technol.*, 62, 9, 1249–1258, 2002. https://doi.org/10.1016/ S0266-3538(02)00063-5.
58. Thompson, D.A. and Best, J.S., *IBM J. Res. Dev.*, 44, 311, 2000.
59. Sun, S.H., *Adv. Mater.*, 18, 393, 2006, doi: 10.1002/ adma.200501464.
60. Collins, P.G. and Avouris, P., *Sci. Am.*, 283, 6, 62–69, 2000.
61. Endo, M., Hayashi, T., Kim, Y.A., Muramatsu, H., *Jpn. J. Appl. Phys.*, 45, 6A, 4883–4892, 2006.
62. Jurewicz, K., Delpeux, S., Bertagna, V., Beguin, F., Frackowiak, E., *Chem. Phys. Lett.*, 347, 36–40, 2001.

3

Nanocomputers in Aid of Defense

Angelica Sylvestris Lopez Rodriguez

Chemical Engineering, Tabasco Juarez Autonomous University, Mexico

Nanotechnology will let us build computers that are incredibly powerful. We'll have more power in the volume of a sugar cube than exists in the entire world today.
Ralph Merkle

3.1 Introduction

In recent years, scientific investigations have focused on the search for methods that minimize the sizes of storage devices and information processing. The reduction of sizes has been a constant effort in the world of computing, which is easily seen when remembering the gigantic computers used a few decades ago and comparing them with the existing minimized computers of today. However, the silicon technology (base of the current microprocessors) is reaching its maximum capacity, which makes it necessary to search for new mechanisms that allow the production of *nanocomputers*. Nanocomputers are formed by several components of nanometer size; although the outside

Madhuri Sharon, et al. Nanotechnology in the Defense Industry: Advances, Innovation, and Practical Applications, (89–108) © 2019 Scrivener Publishing LLC

structure can be of microscopic size. The nanoprocessors would have a power and capacity well above the current microprocessors, with differences so small as to fit on the head of a pin, and therefore would occupy reduced spaces [1].

3.1.1 Classification of Nanocomputers

The nanocomputers fall within the classification of computational nanotechnology, which includes the modeling and simulation of complex structures at the nanometer scale and the management of atoms using nano-manipulators controlled by computers. A nano-manipulator is a specialized microscopic and nanorobotic viewing system that helps scientists and researchers working on extremely small objects. The system was initially designed for microscopic manipulation by computer integration manufacturers. Within the classification of computational nanotechnology are four possible developments of nanocomputers [2]: (i) Electronic, (ii) Mechanical, (iii) Chemical & Biochemical and (iv) Quantum.

3.1.1.1 Electronic Nanocomputers

Electronic nanocomputers contain molecular scale components, that are likely to be up to 10,000 times more densely integrated than today's smallest microcomputers.

Electronic technology is one of several alternative technologies that have been proposed for implementing a nanocomputer. Electronic nanocomputers could be of a magnitude faster than current electronic computers, as well as many times smaller or more densely integrated. Although some of the operating principles for electronic nanocomputers could be similar to microcomputers, there is a limit as to how far the designs and fabrication technologies for microcomputers can be scaled down. These devices and designs take advantage of some of the very effects that have been obstacles to making smaller conventional transistors and circuits. Although electronic nanocomputers will not use the traditional concept of transistors for their components, they will still operate by storing information in the positions of electrons. There are several methods of nanoelectronic data storage currently being researched. Among the most promising are single electron transistors and quantum dots [3]. All of these devices function based on the principles of quantum mechanics. The number of electrons can be changed by adjusting electric fields in the area of the dot. Dots range from 30 nm to 1 micron in size and hold anywhere from 0 to 100s of electrons [2].

3.1.1.2 Mechanical Nanocomputers

Mechanical nanocomputers would use tiny moving components called nanogears to encode information. Such a machine is reminiscent of Charles Babbage's analytical engines of the 19th century. For this reason, mechanical nanocomputer technology has sparked controversy; some researchers consider it unworkable. Eric Drexler and Ralph Merkle are the leading nanotech pioneers involved with mechanical nanocomputers. They believe that through a process known as mechanosynthesis, or mechanical positioning, these tiny machines would be able to be assembled [4].

The fabrication of nanomechanical devices requires some handmade parts. It becomes a tedious job to move a few atoms from one place to another, and it is really a difficult task to manufacture a reliable system using this technique.

3.1.1.3 Chemical and Biochemical Nanocomputers

Chemical and biochemical computers would store and process information in terms of chemical structures and interactions.

In general terms, a chemical computer is one that processes information by making and breaking chemical bonds and stores logic states or information in the resulting chemical (i.e., molecular) structures. In a chemical nanocomputer computing is based on chemical reactions (bond breaking and forming) and the inputs are encoded in the molecular structure of the reactants and outputs can be extracted from the structure of the products, meaning that in these computers the interaction between different chemicals and their structures is used to store and process information.

Biochemical nanocomputers already exist in nature; they are manifested in all living things. But these systems are largely uncontrollable by humans and make artificial fabrication or implementation of this category of "natural" biochemically-based computers a far-off possibility, because the mechanisms for animal brains and nervous systems is still poorly understood. We cannot, for example, program a tree to calculate the digits of π (pi), or program an antibody to fight a particular disease (although medical science has come closer to this idea in the formulation of vaccines, antibiotics, and antiviral medications.

3.1.1.4 *Quantum Nanocomputers*

A quantum nanocomputer stores data in the form of atomic quantum states or spin. Today's computer data is in bits, using a binary system in which every bit has a value of either 1 or 0. Quantum computing exponentially expands the potential of computing by applying quantum mechanics, and measuring data in qubits instead of bits. Technology of this kind of nanocomputer is already under development in the form of single-electron memory (SEM) and quantum dots. The energy state of an electron within an atom, represented by the electron energy level or shell, can theoretically represent one, two, four, eight, or even 16 bits of data. The main problem with this technology is instability. Instantaneous electron energy states are difficult to predict and even more difficult to control.

An electron can easily fall to a lower energy state, emitting a photon; contrarily, a photon striking an atom can cause one of its electrons to jump to a higher energy state [5].

3.1.1.5 DNA Nanocomputers

DNA computing is a branch of computing which uses DNA, biochemistry and molecular biology hardware, instead of the traditional silicon-based computer technologies. In 1994, Leonard Adleman took a giant step towards a different kind of chemical or artificial biochemical computer when he used fragments of DNA to compute the solution to a complex graph theory problem [6]. Like a computer with many processors, this type of DNA computer is able to consider many solutions to a problem simultaneously. The first prototype of a DNA computer, the TT-100 by Leonard Adleman, was a test tube filled with 100 microliters of a DNA solution. He managed to solve, for example, an instance of the directed Hamiltonian path problem [7].

These computers use DNA to store information and perform complex calculations. DNA has a vast amount of storage capacity that enables it to hold the complex blueprints of living organisms. A single gram of DNA can hold as much information as one trillion compact discs [8].

3.2 History of Nanocomputers

In 1959, the physicist Richard Feynman was talking about a future in which tiny machines could develop great works. This was the first time that the technology once considered science fiction was discussed as a reality. Feynman observed that it was possible to build and operate submicroscopic machinery. He proposed that large numbers of completely identical devices might be assembled by manipulating atoms one at a time. Feynman's proposal sparked an initial flurry of interest. However, it did not broadly capture the imagination of the technical community or public. At the time, building structures one atom at a time seemed out of reach [2]. In 1959, it was only possible to put one transistor on an integrated circuit [9]. Twenty years later, circuits with a few thousand transistors were common place [3].

In the 1960s, Gordon Moore, the co-founder of Fairchild Semiconductor and CEO of Intel, observed that every two years there was a doubling of the number of transistors in a microprocessor. This empirical trend has continued, and the principle has come to be called "Moore's Law" [10].

Aviram and Ratner [11] published the construction of a very simple electronic device, a rectifier, based on the use of a single organic molecule. The molecular rectifier consists of a donor pi system and an acceptor pi system, separated by a sigma-bonded (methylene) tunneling bridge.

In the early 1980s, biochemists and geneticists discovered how to splice short lengths of deoxyribonucleic acid (DNA) and ribonucleic acid (RNA) into much longer sequences, leading to the development of the molecular genetics industry. An example of this is the discovery of a biochemical process known as the polymerase chain reaction (PCR) by Mullis [12]. This process allowed the exponential duplication of strands of DNA, amplifying a few molecules of genetic material into macroscopically measurable quantities [13].

By the middle of the1980s, computer miniaturization was enough to fit one million transistors on a chip of no more than 1 cm^2. Also, quantum dots were discovered in the early 1980s. Quantum effects were being harnessed in the development of scanning tunneling electron microscopes (STMs) and atomic force microscopes (AFMs) through which scientists could view and manipulate individual atoms [14].

In accordance with Feynman's vision, these advances of molecular scale were the beginning of Nanotechnology. This term was invented by Taniguchi in 1974, but it was first popularized in the 1980s by the scientist and visionary K. Eric Drexler in his book *Engines of Creation: The Coming Era of Nanotechnology* [4].

In the 1990s, the convergence of developments in physics, chemistry, biochemistry, electrical engineering, and computer science were beginning to form a route towards a practical, useful nanotechnology. A revolution was occurring in miniaturization. The engineering and manufacture of micron-scale machinery has become an industry upon which further nanometer-scale miniaturization could be based. Angell *et al.* [15] explored the manufacturing techniques, properties and applications of silicon micromechanical devices. Figure 3.1 shows a timeline of key events in the development of nanotechnology.

Based on the intuition of Feynman, the team of Gina Adam and Dmitri Strukov from the University of California have designed a functional computing device of nanometer size [16]. The concept is based on a three-dimensional dense circuit (with vertical architecture, unlike the essentially horizontal

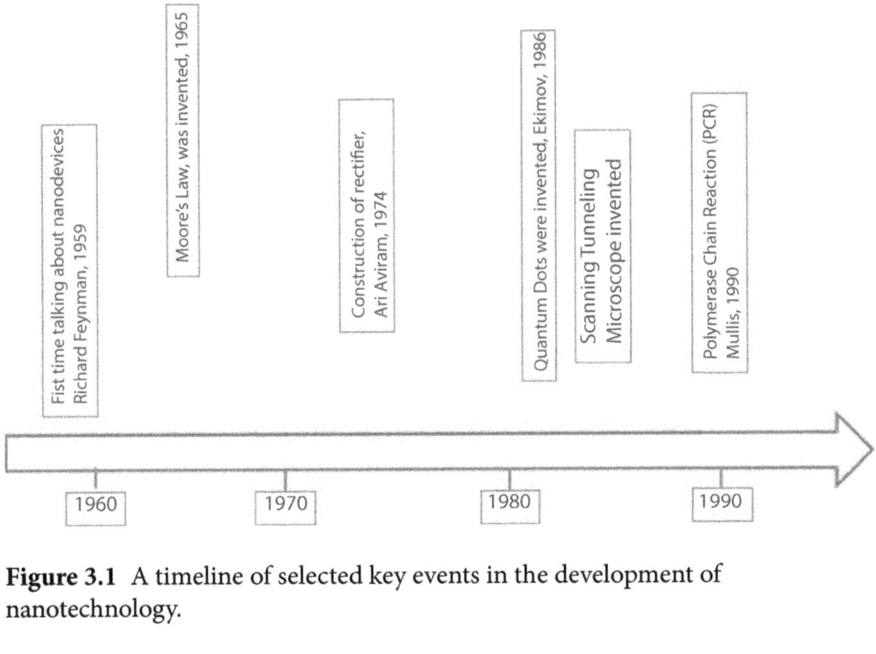

Figure 3.1 A timeline of selected key events in the development of nanotechnology.

architecture of the circuit boards) that operates on a type of nonconventional logic. The nanocomputer could, in theory, occupy a space not larger than 50×50×50 nanometers. An essential resource for this singular computer is the use of memristors, which are circuit elements whose resistance depends on the loads and most recent addresses of the currents that have flowed through them (Figure 3.2).

The nanocomputers work on the logical operations and the storage of information in a simultaneous and local way. This greatly reduces the need for components and space for employees to carry out logical operations and move data between the sites where it is stored and the site where it is processed. The result of the computation is stored immediately in a memory element, which prevents the loss of data in case of power outages, a critical function in autonomous systems such as robots. In addition, researchers have redesigned the traditional two-dimensional architecture of the memristor, turning it into a three-dimensional block, which could then be stacked and bundled with other components in that space of 50 nanometers on each side [16].

The history of computing has involved a sequence of changes from gears to relays, transistors and valves to integrated circuits,

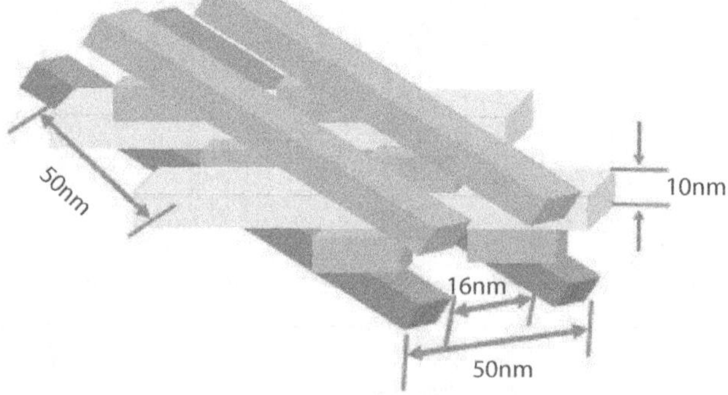

Figure 3.2 Schematic of memristor stacked in a set whose dimensions could fulfill the conditions imposed by the great challenge of Feynman [16].

and so on. The techniques of today can fit logic gates and wires to a fraction of a micron wide on a silicon chip. Soon the parties will become smaller and smaller until consisting of only a handful of atoms. At this point the laws of classical physics break down and the rules of quantum mechanics, so that the new quantum technology should replace and/or supplement what we currently have. Support will be provided to a completely new type of computing with new algorithms based on quantum principles [1].

Different computer manufacturers have made significant progress in this field, including Hewlett-Packard, which, together with scientists from the University of California, have patented a process for the production of nanocomputers. The patent covers a process that brings together several different functions in a single nanoprocessor. This is possible thanks to the division of the processor in a number of areas that can perform independent calculations.

3.3 The Nanocomputers

One disadvantage of Moore's Law is the scale. As previously mentioned, the principle of this law is that every two years there is doubling in the number of transistors in a microprocessor. The latest chips from Intel use silicon transistors with features smaller than 14 nanometers [1]. According to the advances of technology, by 2020 the industry will need to be down to five nanometers; this behavior is opposite to Moore's Law, opening up the knowledge of quantum physics. An alternative is to change the actual transistors of silicon and build computers with transistors made from carbon nanotubes (CNTs). In 2013, a team of researchers guided by Prof. Subhasish Mitra at Stanford University, built the world's first computer prototype based entirely on CNTs. They named it Cedric. The CNT computers may be far more energy efficient than traditional silicon-based systems. Moreover, they can improve the speed of computers, reducing heating, because this kind of computer has the ability to dissipate heat [17].

Another study was carried out by scientists at Tsinghua University in China. They fed silkworms with a solution containing CNTs and graphene. They observed that these silkworms produced a super silk that was not only incredibly strong but was also able to conduct electricity. According to the researchers, the discovery could be carried out with very resistant, biodegradable protective fabrics, medical implants, and portable and ecological electronics.

Needless to say, nanocomputers will have great impact on all the gadgets used in defense, be it armor, sensor, stealth, submarine, communication system, healthcare, or warfare.

3.3.1 Nanotechnology and Quantum Computers

The information of current computers is represented by bit sequences. These sequences are introduced in processors to perform certain operations with them (add and subtract, display images on a screen, play music, search for an address, etc.). The classical computing is binary and sequential; it is binary because the processor uses bits that can take two values (0 and 1); and is sequential because the operations performed by the processor run one after the other. These operations are carried out by components as transistors, capacitors, connectors, etc. [18].

The miniaturization of electronic devices sooner or later will come to end, when the laws of physics prevent us from further decreasing the size of circuits. At any given time, we will not be able to continue with the current silicon processors, binaries and sequential. One of the most viable ways is to build computers operating with radical difference: the quantum mechanics. The arrival of quantum computers will involve a radical change in the hardware and software that we now know but will allow one to perform calculations that are unthinkable today.

One of the most important scientists in this field is the Spanish physicist Juan Ignacio Cirac, the 2006 Prince of Asturias Award winner, who has been the Director at the Max-Planck Institute of Quantum Optics, a renowned center of German research, since 2001 [19].

3.3.2 Recent Advances in Nanocomputers

Scientists from the Hewlett-Packard laboratories in Palo Alto, California, and the University of California, Los Angeles (UCLA), are developing very small computers, i.e., as small as a grain of sand. These new computers could be compared with molecules.

How can the computers be so tiny? All computers are based on an on-off switch. Scientists have developed a rotaxane which acts as an on-off switch. A rotaxane is a mechanically intertwined molecular architecture consisting of a "dumbbell-shaped molecule," which is inserted through a "Macrocycle" [20]. The rotaxane is inserted between two crossed wires. When the molecule is in "off " position, an electron can jump from a cable up to the molecule and then from there to the other cable. Rotaxane works as an electronic transistor. Figure 3.3 shows a diagrammatic representation of a rotaxane.

The scientists plan to introduce layers of molecules of rotaxane in the interior of ultra-powerful computers. The new computers will be smaller, more economical and 100 trillion times

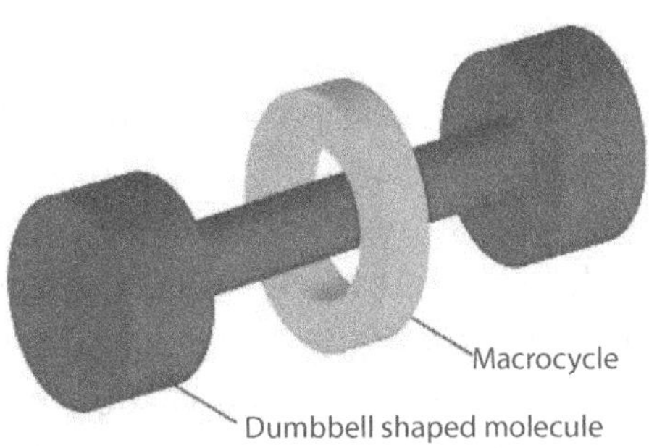

Figure 3.3 Schematic of a rotaxane.

faster than current computers. They are called chemically assembled electronic nanocomputers (CAEN). One disadvantage of rotaxane is that it can only be used once. For this reason, it serves only to store information in a read-only memory.

With Moore's Law quickly edging towards physical limitations, faster computers will soon require new technologies at small scales. Nowadays, the general idea of a nanocomputer involves the insertion of a small device inside the body to maintain natural organic functions for the treatment of diseases or any other activities that the body is not able to develop [21].

3.4 Applications of Nanocomputers in the Military

The implementation of nanotechnology causes a drastic reduction in energy consumption and dramatically advances the ability of medicines to cure and prevent diseases, and also to add significant increases in the precision and effectiveness of military devices and weapons [22]. The use of nanocomputers in war could create more resistant materials like bullets, armor, and other devices. Nanotechnology has potential applications for defense, especially in the field of sensors, transducers, nanorobotics, nanoelectronics, memory storage, propellants, and explosives to enhance the performance of devices and weapon systems [23]. The broad uses of nanotechnology for defense applications are given below in Figure 3.4.

Nanotechnology is being applied in armament electronics where miniature circuit building, actuation, and detonation systems are involved. Nanoelectronic components are now manufactured by hybridization of silicon components, where single logic circuits are made out of self-assembled single molecule transistors. Nanowires capable of conducting electricity and various new forms of memories and storage devices have become feasible. One of the novel examples is use of CNTs.

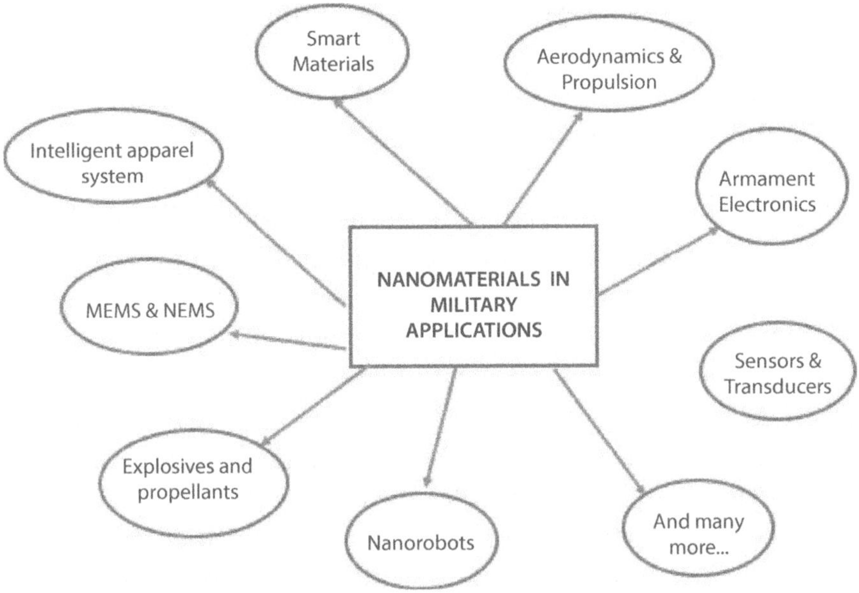

Figure 3.4 Applications of nanocomputers in defense [23].

Their special features are almost 50,000 times thinner compared to human hair, as strong as diamond, and have the ability to hold 100 times the current of metal wires. The reduction in size of the systems from computers to wireless phones is a continuing trend for electronic defense systems. The significance of this miniaturization goes well beyond just smaller size and reduced weight. Batch fabrication and combination of components and subsystems into fewer chips has been a key driver in the miniaturization of microelectronics, enabling reduced cost, increased reliability, and robustness through the parallel manufacturing of many integrated components, and increased functionality in ever-smaller packages. A robot typically requires sensors for pressure, position, vision, temperature and movement, and a computer to process signals and decide the action controls.

Nanotechnology applications will likely pervade all areas of the military. Very small electronics and computers will be used everywhere, e.g., in glasses, uniforms, and munitions.

Sensors play critical roles in the functioning of tiny electronic circuits for large and complex systems. For instance, electronic and optoelectronic circuits need voltage, current, temperature, light, and other sensors to operate whereas jumbo jets need mechanical sensors and actuators to function.

CNTs have been realized as chemical sensors capable of detecting small concentrations of toxic gas molecules such as NO_2, NH_3, etc. The advantage of CNT-based chemical sensors over conventional gas sensors are: mild reaction conditions (functions at room temperature), fast response, higher sensitivity and high adsorption surface area. Nanotube-based gas sensors are used for detecting biochemical weapons, land mines, air pollution and even organic molecules in space [24].

Nanocomputers working with 16 KB memory will be 10 microns by 10 microns or about 30 to 100 times the density planned by the computer industry for their memories. For miniature-scale robots, it is anticipated that by using nanotechnology, a mini robot the size of a housefly is likely to be developed, which will be driven by a nanocomputer [23].

Nanostructured materials will provide the defense industry with the required means to develop lighter, flexible, more agile and more resistant military platforms, including light armored vehicles, tanks, fighter jets, man-transportable micro unmanned air vehicles (MUAV), etc. Developments in NT will ensure the manufacture of platforms that will have an ability to adapt to all types of climate conditions and environments [25]. For example, it is more than plausible that, in the future, NT will assist in the production of aerial platforms able to operate from land bases or sea-based carriers, regardless of the environmental constraints (maritime spindrifts, desert sand, tropical vegetation and extreme level of wetness).

Remarkable mid-term advances are expected to occur in information technologies (IT). Battle system architectures, lightweight nano-networks, and self-assembled nanosystems will noticeably improve situational awareness capabilities.

When preparing a campaign or a crisis intervention, powerful and highly capable computers will support the commander to assess data issued from various sensors should they be located on earth, sea, air or space. Network-centric warfare concepts will undoubtedly benefit from such advancements in computing speed. Whether such capabilities will definitely erase the "fog of war" remains uncertain.

3.5 More Powerful Computers to Come

Large-scale systems are expected to be developed for strategy planning, battle management and logistics. With nanotechnology-improved computers, it is more than probable to assist in a drastic reduction in size of the various components and armament subsystems.

Such an evolution could lead to the development of armaments that will be characterized by a higher lethal payload. It must be added that computers will, at the same time, become much faster and far less power-consuming (thus maybe reducing the operational tyranny of the logistic chain during military campaigns). In a longer-term perspective, one cannot exclude the development of systems augmented by a new level of artificial intelligence (AI) throughout the military. Such systems could well be embedded in all the equipment of soldiers (rifles, glasses, uniforms, ammunitions, micro-robots, nanobots, etc.) and invade a soldier's physiology.

On a strategic level, nano-enabled systems could reveal new potentials for planners and battle management services. Together with sensors (via smart dusts?), wireless communication components, lightweight displays, and nanotechnology may well lead to the instauration of a global, if not ubiquitous, network.

One of the first applications of nanotechnological solutions for soldiers lies in the development of ambient-intelligent networks (AIN). An example of application of AIN is human fitness and health monitoring in which small devices, such as cell phones or receivers, become accessible for various heart rate and blood monitors or calorie counters. The main advantage of AIN resides in the fact that they rely on noninvasive technologies. Otherwise, soldier-worn systems could be able to sense the state of health of the wearer and could generate rapid reactions by releasing drugs or by using small materials in order to compress the wounds. Such invasive nano-agents could also help the soldier to progress on the battlefield, should the combat environment be infected by biological, bacteriological or chemical agents. Nanostructured alert systems could trigger the rapid delivery of drugs and other medicines until search-and-rescue gear can extract the injured soldiers from the infected zone of the battlefield.

Electronics and computers will become much smaller and at the same time much faster and less power-consuming through the use of nanotechnology and especially nanocomputers. Such systems, augmented by new levels of artificial intelligence, will be used throughout the military, even embedded in very small components (rifles, glasses, uniforms, mini- and micro-robots, ammunitions). On the other hand, large battle-management and strategy-planning systems will encompass many layers and a high degree of autonomous decision-making. Together with sensors, wireless communication components and small, lightweight displays, they would form a ubiquitous network not only on the battlefield, but also, for example, in logistics.

Actually, new research is focused on studying new molecular and supramolecular materials with chemical, physical, electrical, magnetic and optical properties for use as nano-chemical devices. These have application in molecular electronics or biomedicine as machines at the atomic scale with cleaning functions of arteries damaged by arteriosclerosis, DNA repairers or re-builders of cells. These materials are rotaxanes, which were discussed above.

When incorporated into the bloodstream of a person, for example, these microscopic computers could identify bacteria that are not older than them, and decide the specific drugs to be delivered to combat the infections.

The first impulse to use electronic circuits emerged from the ballistic missile programs, because of their limited load capacity. The microtechnology was gradually developed and applied immediately to computers, which greatly reduced the size of its processors. And now nanotechnology is taking over the role of developing a much-reduced size of computer having much higher multidirectional performance activity. It will be worth mentioning here that due to the secretive and proprietary nature of such development, it is not possible to give a real account of the development that has occurred.

3.6 Summary

This chapter has discussed the basic concepts of computers. Every technology needs advancements, and one of the major requirements in advancing computers are their miniaturization, which is going to be of great help in every aspect of defense activity. Nanotechnology is of great help in not only miniaturization but also in developing more powerful nanocomputers. Efforts in this direction have also been touched upon in this chapter.

References

1. Hirwani, D. and Sharma, A., *Recent Res. Sci. Technol.*, 4, 3, 16–17, 2012.
2. Montemerlo, M.S., Love, J.C., Opiteck, G.J., Goldhaber, D.J., Ellenbogen, J.C., *Technologies and Designs for Electronic*

Nanocomputers, pp. 1–6, The MITRE Corporation, Mc Lean, Virginia, 1996.

3. Turton, R., *The Quantum Dot: A Journey into the Future of Microelectronics*, Oxford University Press, Oxford, U.K, 1995.
4. Drexler, K.E., *Engines of Creation: The Coming Era of Nanotechnology*, Anchor Books, New York, 1986.
5. Sahni, V. and Goswami, D., *Nanocomputing: The Future of Computing*, pp. 1–23, Tata McGraw-Hill Publishing Company Ltd, New Delhi, 2008.
6. Adleman, L.M., Computing with DNA: The manipulation of DNA to solve mathematical problems is redefining what is meant by "computation. *Sci. Am.*, 94, 54–61, 1998.
7. Ogihara, M. and Ray, N., *Nature*, 403, 13, 143–144, 2000.
8. Braich, R.S., Johnson, C., Rothemund, P.W.K., Hwang, D., Chelyapov, N., Adleman, L.M., Solution of a satisfiability problem on a gel-based DNA computer, in: *DNA Computing. DNA 2000. Lecture Notes in Computer Science*, vol. 2054, A. Condon and G. Rozenberg (Eds.), pp. 27–42, Springer Berlin, Heidelberg, 2001.
9. Meindl, J.D., *Sci. Am.*, 257, 4, 78–89, 1987.
10. Wu, J., Shen, Y.-L., Reinhardt, K., Szu, H., Dong, B., *Appl. Comput. Intell. Soft Comput. Arch.*, 2013, 1–13, 2013.
11. Aviram, A. and Ratner, M.A., *Chem. Phys. Lett.*, 29, 2, 277–283, 1974.
12. Mullis, K.B., The unusual origin of the polymerase chain reaction. *Sci. Am.*, 262(4): 56–65, 1990.
13. Ma, T.S., *Chest*, 108, 1393–1404, 1995.
14. Hansma, H.G. and Lía, P., *Curr. Opin. Chem. Biol.*, 2, 579–584, 1998.
15. Angell, J.B., Barth, P.W., Terry, S.C., *Sci. Am.*, 248, 44–55, 1983.
16. Adam, G.C., Hoskins, B.D., Prezioso, M., Merrikh-Bayat, F., Chakrabarti, B., Strukov, D.B., *IEEE Trans. Electron Devices*, 64, 1, 312–318, 2017.
17. Venkatesh, M. and Zhou, Ng A., *Proceedings of the 2009 International Conference on Software Technology and Engineering*, World Scientific, pp. 86–88, 2009.

18. Gennady, P.B., Doolen, G.D., Mainieri, R., Tsifrinovich, V.I., *Introduction to Quantum Computers*, pp. 1–7, World Scientific Publishing Co. Pte. Ltd., Singapore, 1998.
19. Waldner, J.B., *Nanocomputers and Swarm Intelligence*, pp. 89–130, John Wiley & Sons, Inc, 2008.
20. Sauvage, J.P. and Dietrich- Buchecker, C., *Molecular Catenanes, Rotaxanes and Knots: A Journey Through the World of Molecular Topology*, pp. 1–6, Wiley-VCH, D-69469 Weinheim, 2008.
21. Karkare, M., *Nanotechnology: Fundamentals and Applications*, I.K. International Publishing House, New Delhi, 2008.
22. Patil, M., Mehta, D.S., Guvva, S., Future impact of nanotechnology on medicine and dentistry. *J. Indian. Soc. Periodontol.*, 12, 2, 34–40, 2008.
23. Kharat, D.K., Muthurajan, H., Praveenkumar, B., *Synth. React. Inorg. Met.-Org. Nano-Met. Chem.*, 36, 231–235, 2006.
24. Misra, A., *Curr. Sci.*, 107, 3, 419–429, 2014.
25. De Neve, M.A., Royal Higher Institute for Defense Center for Security and Defense Studies. *Focus Pap.*, 8, 1–13, 2009.

4
Nanotechnology-Aided Armor

Pio Sifullentes Gallardo

Chemical Engineering Faculty, Juárez Autonomous University of Tabasco, Mexico

Things on a very small scale behave like nothing that you have any direct experience about. They do not behave like waves, they do not behave like particles, they do not behave like clouds, or billiard balls, or weights on springs, or like anything that you have ever seen.
Richard P. Feynman

4.1 Historical Background of Armor

Armored materials serve to protect life. In ancient times, armored materials were made up of raw materials of natural origin such as wood and stones. Then skins of animals and metals were used. These materials that we now know as ceramics, metals, biomaterials, polymers and others have undergone man-made transformations; these changes have usually been made at the macro and micrometric scale [1]. The first patent registrations that led to a change in technology appeared in 1919, referring to the first bulletproof vests, which include great advancements in the knowledge of natural fibers. Then came the technology of plastics or ballistic polymers such as

Madhuri Sharon, et al. Nanotechnology in the Defense Industry: Advances, Innovation, and Practical Applications, (109–132) © 2019 Scrivener Publishing LLC

Nylon, with more technological advancements coming in the 1960s, when the first body armor was built with a new generation of high-performance fibers. The current requirement is to design armor materials at a smaller level called nanometric, which incorporates the inclusion of composite materials whose design, research and manufacturing is necessarily multidisciplinary.

It is now known that to protect the lives of human beings it is not only necessary that the armored materials have resistance to being penetrated by projectiles or bullets, but also by sound waves (which can be life-threatening), radiation, viruses or dangerous substances, commonly gases.

On the other hand, it is important to highlight that there are properties of intelligent materials such as self-repair, camouflage, memory and the delivery of medicines by fibers or materials that are being incorporated into military technology.

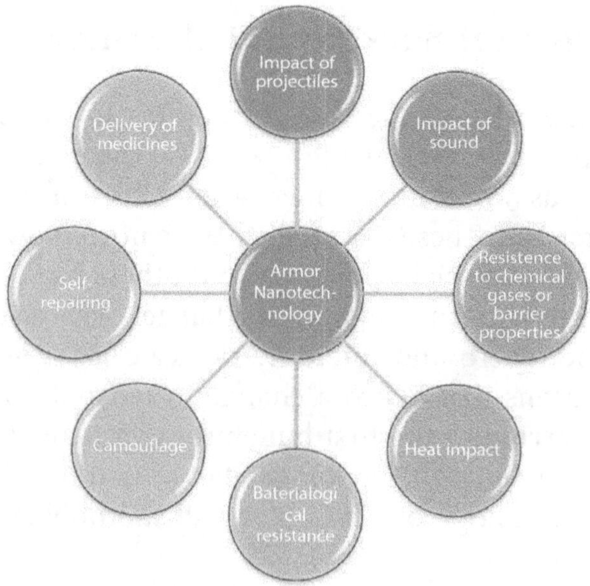

Figure 4.1 Some characteristics of armor nanomaterials.

The characteristics of armored materials which function better with nanotechnology are: (a) reduction of projectile impact, (b) resistance to chemical gases or barrier properties, (c) reduction of heat impact, (d) reduction of sound impact, (e) bacteriological resistance, (f) camouflage, (g) ability to self-repair and (h) ability to deliver medicines (Figure 4.1).

Therefore, it is possible to define armored nanomaterials as the branch of science that studies the structure and properties of materials that serve the preservation of life focused on its nanometric scale.

4.2 Nanomaterial-Aided Armor

For developing an armor for the all-round safety of defense personnel, the nanomaterials that are required encompass polymers, carbon nanoforms, nanocomposites, nanomaterials and biomaterials for smart armor. They are briefly discussed below.

4.2.1 Polymers

Plastics and resins are also known as polymers. Their chemical structure is composed of many repeating units that are linked to form a substance of high molecular mass. There are many types of polymers, most of which are currently in the research phase. It is a very flexible area where molecules can be designed on a nanometric scale. Although nature has been able to manufacture materials or polymers that have great impact resistance, such as silk, in laboratories it is also possible to synthetically design these materials. One of the first ballistic materials was nylon, followed by polyaramides, such as Kevlar, which is mostly used today, ultrahigh molecular weight polyethylene and polycarbonate. In order to manufacture armored polymeric materials, chemical modifications can be done during the polymerization reactions and physical transformations during fiber production, both processes are explained below.

4.2.1.1 Polymerization Reaction

The design of polymers during the chemical reaction implies the use of nanometric technologies in the areas such as (a) nanoreactors (b) microreactors, (c) catalysts, (d) molecular structure design and reagents and (e) molecular mass.

a. ***Nanoreactors*** – These reactors are nanocontainers where chemical reactions are performed and have existed for a long time in nature as nuclei of cells, some pores that are formed in protein channels in addition to some particles synthesized in the laboratory [2]. An example of these reactors is shown in Figure 4.2. An example of a polymerization nanoreactor that occurs in nature is the capillary conduits of the *Hevea brasiliensis* tree (rubber tree), which produces natural rubber latex (cis-1,4-polyisoprene) with particle size of a few nanometers.

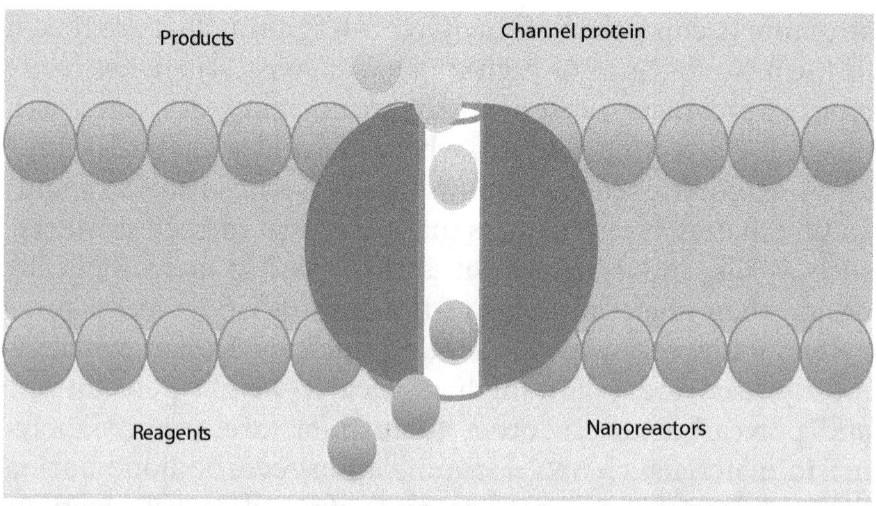

Figure 4.2 Nanochannels in proteins used as nanoreactors.

Sometimes it is possible to manufacture nanoreactors using some viruses or polymeric molecules [3, 4]. Still under discussion is the nanofabrication of small containers where one or a few molecules can be introduced to make them react, which opens the possibility of studying new phenomena such as introducing molecules of the size of the nanoreactors and observing the phenomena that occur as well as the properties of the obtained products [5].

An important area for nanoreactor design has been developed in a very outstanding way with the manufacture of nanotubes, because they have channels to use as nanoreactors. We find that not only lipid nanotubes, but also nanotubes, can be synthesized in the laboratory from titanium dioxide, zinc oxide and carbon, the latter being discovered by Ijima, which have special characteristics because they can be manufactured in a very versatile way as single-wall or multiple-wall CNTs, depending on the synthesis method, and they could be used for reactions where greater mechanical resistance is required. Moreover, the diameter can be controlled using different diameters in catalyst particle. The proposed mechanism to obtain carbon nanotubes is shown in Figure 4.3 [6].

One of the advantages of using nanoreactors is that one can know reactions taking place between free molecules and the effects of spherical interactions of molecules, which could help to design new materials; however, they have been little used to obtain armored materials.

b. ***Microreactors*** – Performing chemical reactions in microreactors is more common than in

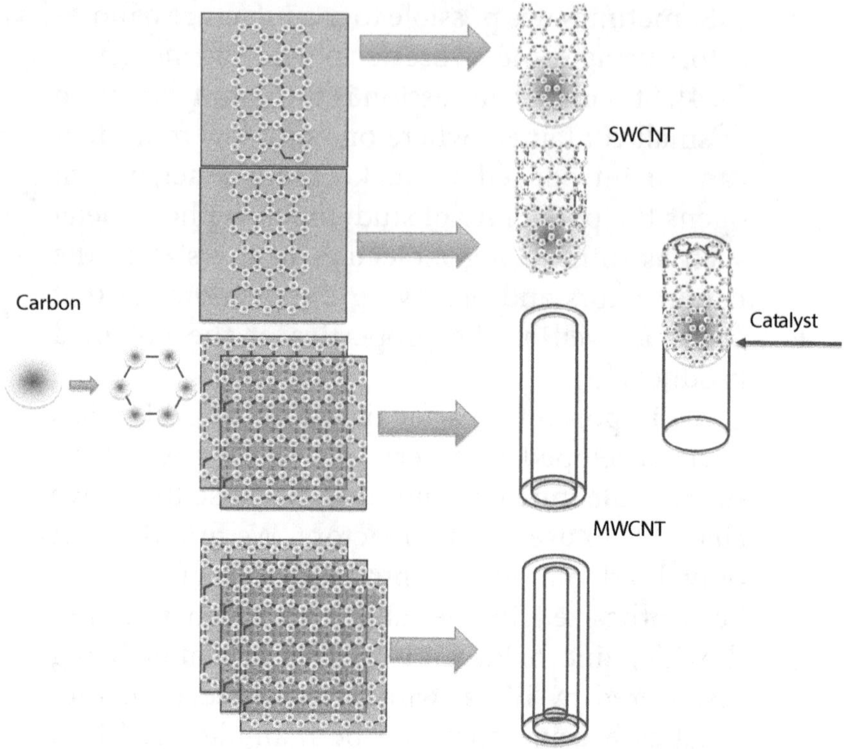

Figure 4.3 Fabrication of SWCNTs and MWCNT on metal particle surface.

nanoreactors. Microreactors are defined as containers in which a manufacturing technique using microtechnology is used partially or completely. These reactors are easy to design, build and operate. The complete system includes micro-mixer, micro-heat exchangers and microseparators, all of which permit the study and scaling up of new materials at low cost.

c. ***Catalysts*** – These are substances that speed up chemical reactions without taking part in the reaction; but in the case of polymers they can also substantially modify the chemical structure and thereby significantly improve the final properties.

For example, catalyst is used for synthesizing polyethylene of ultrahigh molecular mass for armor in order to obtain polyamides, such as Kevlar, and nano-metal is used as catalyst for the manufacture of CNT, which is a very important armored material due to its great mechanical resistance.

Nanotechnology to obtain very small particles is a process that consists of a series of chemical reactions that decrease the size of the metal catalyst to a few nanometers in diameter. Chemical reactions may occur on the surface of the particle; for example, carbon nanotubes of the same size as the catalyst particle diameter may grow. This technology will be explained in the section on carbon technology of this chapter.

d. ***Molecular Structure Design and Reagents*** – For nano-design or molecular engineering [7], it is necessary to consider the chemical structure of the reagents that will be used to obtain the new nanomaterials. In the case of polymers, it is necessary to consider (a) the selection of aromatic rings that increase the stiffness in the molecular chain of polymer and (b) the size of the molecular mass of the polymer.

The aromatic rings of the reactants will increase the stiffness of the polymer chain; an example of this are the polyaramides, among them Kevlar® that is used in bullet proof (Figure 4.4).

In addition, the electric charges of these groups will give them a high degree of crystallinity, which is necessary to further increase the mechanical strength of the polymer.

e. ***Molecular Mass of Polymer*** – The size of the molecular mass of a polymer is very important because the mechanical properties improve

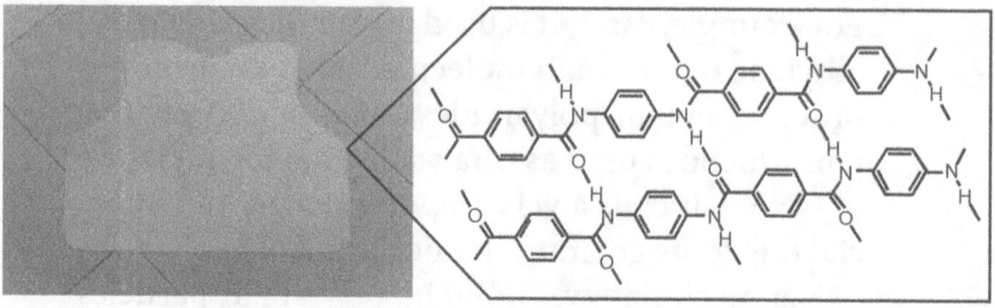

Figure 4.4 A bulletproof vest made of Kevlar fabric and the chemical structure of polyamides.

substantially during the chemical reaction. For this reason, the objective is to increase the molecular mass of the polymers during the chemical reaction, obtaining polymers of ultrahigh molecular mass. For example, low molecular weight polyethylene (LDPE) has a low mechanical strength while ultra-high molecular weight polyethylene (UHMWPE) increases its impact resistance considerably and can be used to make armored polymers. The technology to obtain these polymers is different. For example, in Figure 4.5 the case of polyester is presented, which is realized in two stages in a reactor; although the reactor is a large container the molecular design remains nanometric scale.

The molecular mass of some polymers, such as polyaramides or polyesters, can be increased by carrying out solid-state polymerization (SSP) or swollen-state polymerization (SWSP) reactions. In order to perform these reactions, it is necessary to realize the common polycondensation reactions in which the molten polymer is extruded and cooled in water to obtain pellets or fibers [8].

During the solid-state polymerization process, the pellets are dried under vacuum, then the temperature is increased without generating

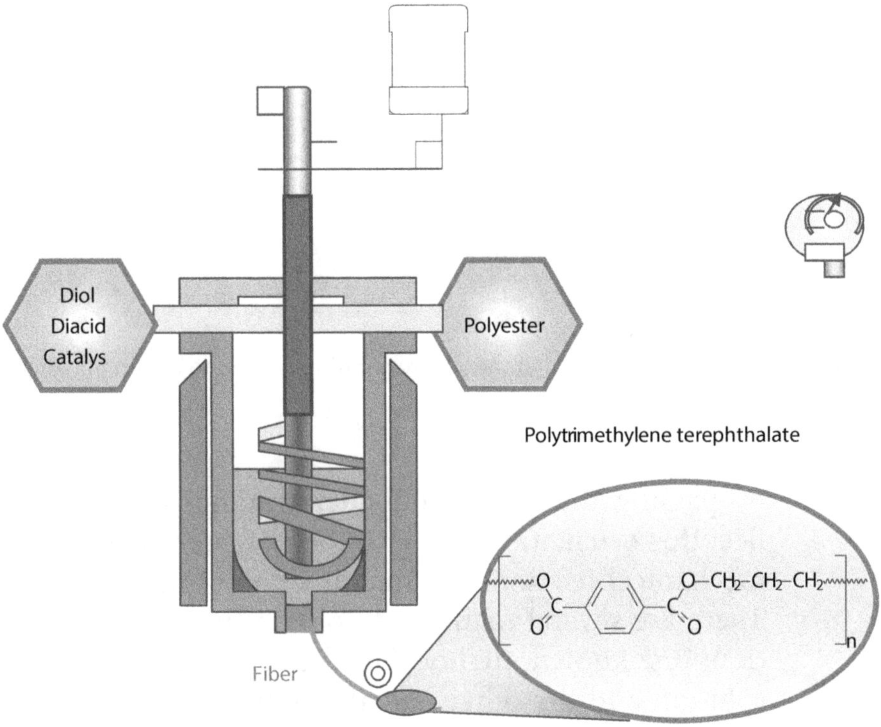

Figure 4.5 Process for obtaining polyesters.

agglomeration. Then, with the help of vacuum extraction only a remaining amount of oligomers are extracted without causing degradation. This process takes a few hours depending on the processing conditions and the polymer used.

In the swollen-state polymerization process, it is necessary to use a solvent that will swell the polymer in the initial stage and then extract the oligomer from inside the pellets. In this process it is also necessary to apply vacuum.

It is possible to modify the process without affecting the chemical structure of the polymer. In this way it is thus possible to achieve the orientation of the polymer chains in the flow direction

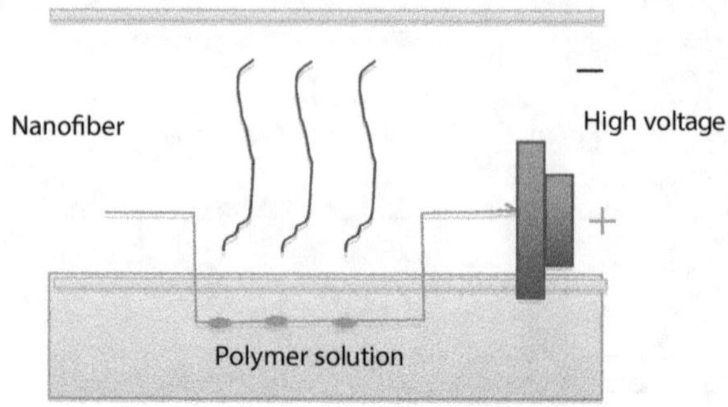

Figure 4.6 Schematic diagram of the electrospinning process to obtain polymer nanofibers.

by passing the molten polymer through an orifice; this orientation will give the polymer greater mechanical resistance in the direction of the flow. There are several methods to obtain this orientation; one known method for fiber extrusion is in solution and the other one also applies magnetic fields to produce nanofibers.

The orientation of the polymer chains can also be achieved by applying an electric field, which will cause the polymer chains to be attracted from the solution connected to one pole to another piece connected to the opposite pole (electrospinning). Through this procedure, it is possible to obtain polymer nanofibers and nanofibers of nanocomposites [9]. This process is called electrospinning, which is shown in Figure 4.6.

4.2.2 Carbon Nanoforms

Although steel is an impact resistant material, other materials have also been discovered that are equally or more resistant and also lighter; of these materials, carbon fibers, which were discovered in the late 1950s, were one of the first known. There are materials obtained from polyacrylonitrile (PAN) that are still used today, but with the discovery of CNTs in 1991 by Professor Iijima, CNT, one of the most impact resistant materials known at present, 100 times stronger than steel, began being used [10, 11]. There are other nanoforms obtained from carbon that can be used for the nano-reinforcement of polymer matrices, such as sheets of graphene or fullerenes [12].

Due to the fact that one of the most important fillers is carbon nanotubes, special emphasis will be given to these particles. The CNTs are formed by carbon atoms and have dimensions less than 10 nm in diameter and are sometimes formed by more than one concentric wall; the former are called single-walled carbon nanotubes (SWCNTs) while the other is referred to as multi-walled carbon nanotubes (MWCNTs). These two types of carbon nanotubes are shown schematically in Figure 4.3.

4.2.2.1 Synthesis of Carbon Nanotubes (CNT)

Some of the common methods used to synthesize CNT are arc discharge, laser ablation and chemical vapor deposition (CVD). In each method, some variants have been developed that allow research and development to improve cost and quality. One of the most used methods is CVD, where equipment consisting of two high temperature furnaces is used; in the first of them a substance or precursor containing carbon is used, which is pyrolyzed to obtain the carbon atoms. The reaction atmosphere in a CVD set-up has to be free of oxygen. The reaction temperature of the CVD furnace depends on the degradation temperature of the precursor used for synthesizing CNT. The carbon atoms are transported using a carrier gas that can be argon, nitrogen or hydrogen. In the second furnace a quartz sample holder is placed, onto which the catalyst composed of nano-metal oxide is deposited. In nano-metal oxide, the catalyst is prepared using a metal salt, which is further reduced with the help of heat and hydrogen to pure metal particles. On these particles the carbon atoms are deposited, which grow on its surface as CNT. A schematic of the equipment used in the process is shown in Figure 4.7.

In this technique, oxides of metals, often iron, nickel and cobalt, are commonly used as catalysts, in which the size of the particle is very important because the diameter of the tubes depends directly on the diameter of the catalyst particles. The catalyst remains inside the nanotubes, and in some cases it is considered as a contaminant, so its purification is necessary (Figure 4.8).

The purification consists of carrying out a reaction with a solution of hydrochloric acid, nitric acid or a mixture of both acids that reacts with the catalyst but does not degrade the carbon nanotubes. Several washes are realized until a neutral pH is achieved; finally, the separation of the carbon nanotubes is obtained by applying centrifugation followed by drying in an oven.

In addition, other residues remain, such as graphite mixed with carbon nanotubes. The carbon nanotubes are thermally stable; the amorphous carbon residues are eliminated by

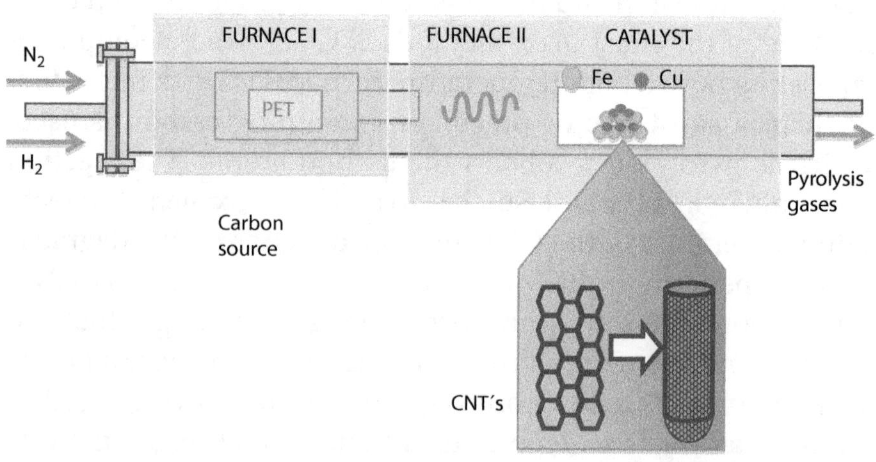

Figure 4.7 A CVD set-up used for synthesis of CNT from a carbon-rich precursor by pyrolysis.

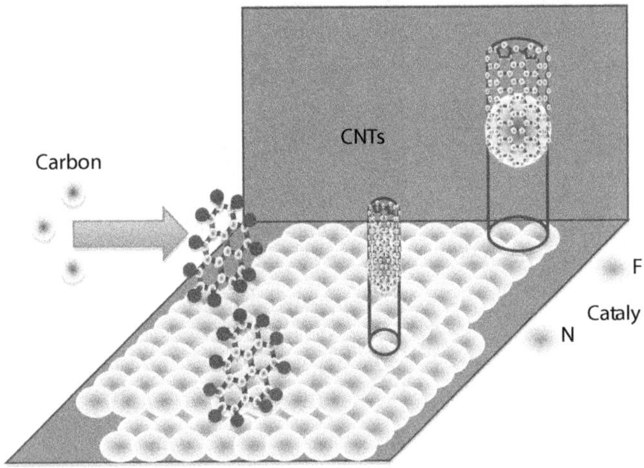

Figure 4.8 Growth of CNTs on the surface of nanocatalyst.

calcination at not higher than 750 °C. In Figure 4.8, the process for obtaining carbon nanotubes is represented schematically.

Although the CVD process often yields a mixture of both SWCNT and MWCNT, it is reported [13] that taking care of the process injecting steam mixed with hydrogen and applying temperatures of 900 °C in the catalyst furnace, as suggested by Iijima, or controlling other parameters during CVD, production of SWCNT can be increased or standardized.

4.2.2.2 *Functionalization of CNT*

For armored nanocomposites it is necessary to have good chemical bonds between the polymeric matrix and the CNTs because the mechanical properties improve substantially, besides improving other properties such as solubility and processability, increasing the number of possible applications.

Functionalization means adding functional chemical groups to the CNT that can react with the matrix to obtain a good chemical bond between the two components. There are several methods for adding functional groups that may be located in open ends and gaps in the walls of CNTs (Figure 4.9). Among

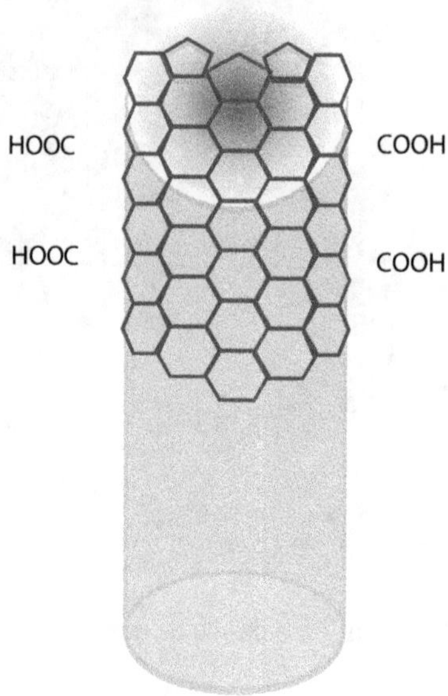

Figure 4.9 Functionalization of CNTs.

the groups added to the carbon nanotubes, amidation, esterification, halogenations, hydrogenation and oxidation have been achieved, among others; the latter because during the purification the temperature can be increased in the presence of oxygen.

4.2.3 Nanocomposites

Nanocomposites are constituted by two or more materials that can be polymers mixed with ceramics or metals. Taking advantage of the individual properties that each of them separately have, they are designed for specific applications; for example, one of these applications is in armored compounds. The concept of nanocomposites applies if nanofillers are used in a continuous matrix, as can be seen in Figure 4.10. Several fillers or polymeric matrices can be used depending on the requirements. Moreover, it is recommended to consider the nanomatrix and nanofillers properties separately prior to use, such as interface or interaction between each of the components to achieve

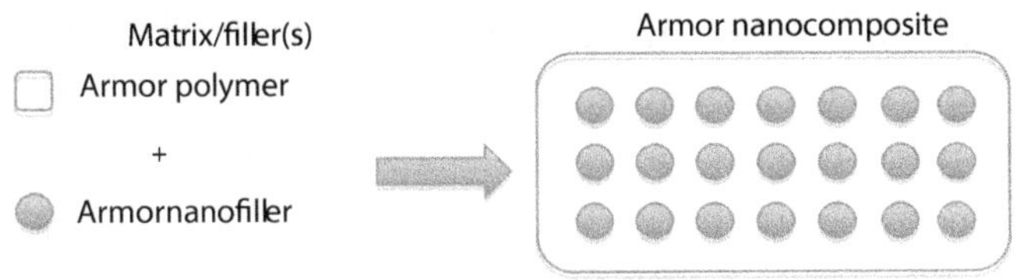

Figure 4.10 Nanocomposites used for fabricating armor material.

bonds between them, so that they reinforce the mechanical properties.

In Figure 4.11 a simplified example of the design of armored nanocomposites is presented in a schematic way. It is important to mention that normally more properties are required to obtain armor nanocomposites and therefore a greater number of materials. In the previous example, the polymer must also have the property of flexibility in addition to ballistic properties, as in the case of polyaramides. In the manufacturing area, the filler is known as reinforcement of the polymer. An example of fillers are carbon fibers in the polymer matrix; the fragility of the carbon fibers is reduced with the flexibility of the polymer, obtaining two characteristics in a single material.

The properties of nanocomposites are different from those of composites; for example, smaller amounts of filler are used to obtain the same mechanical properties, saving significant costs. Another characteristic is that due to the great interaction of the filler nanoparticles, the viscosity increases

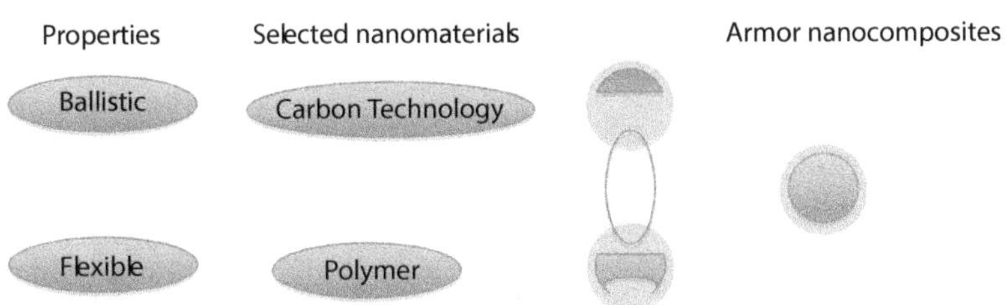

Figure 4.11 Simplified example of armor nanocomposite design.

drastically; therefore, the corresponding modifications have to be made to the process equipment for obtaining fibers or plates, and due to its high viscosity it is more difficult to extrude.

Hazell has used polymers and metals [14] to synthesize nanocomposites. For armored materials the common metals that are used include steel, aluminum, titanium and boron (metalloid). In the case of ceramic materials, glass fiber, alumina, silicon carbide, boron carbide, carbon fibers, graphite and more recently CNT immersed in polymeric matrices have been used [15–17].

Among the materials, there is a greater tendency to use carbon because it is a light and very resistant material when mixed or bonded with polymers, normally polyaramides or polymers of high molecular mass, as can be seen in Table 4.1 [18]. It can be seen that the lightest and most resistant material is the carbon nanotubes, besides that it is a thermally stable material since its degradation occurs until after 750 °C in air. As it can be seen in the Table 4.1, steel is not written because it is a heavy material, although at present it is still used because it is a very economical material and has advantages because it is very easy to machine and weld.

Table 4.1 Materials with high mechanical strength properties.

Fiber	Density (g/cc)	Tensile strength (GPa)	Modulus (GPa)
Polyaramids	1.44	2.8	70–170
Carbon	1.66	2.4–3.1	120–170
Boro	2.5	3.5	400
CNTs	1.33	Up to 50	Up to 1000

4.2.3.1 Processes for Preparing Nanocomposites for Armor

Melt Processing — In this method the polymer is melted along with filler by applying heat. It is a very common method used by the industry because it is a well-known and easy process; however, it is not convenient to melt all polymers, especially high molecular weight polymers and polyaramides that have rings in their structure that give them exceptional stiffness, and therefore this method has some limitations.

Solution Mixing — In this method the polymer is dissolved in an appropriate solvent, then the armor filler like CNTs are added and intensive mixing is done; sometimes ultrasound techniques are applied to separate the agglomerated carbon nanotubes. Sometimes a mixture of solvents is used and it is also necessary to add heat during stirring to dissolve the polymers. When the particles are very small the surface area increases considerably and therefore there is a greater interaction that makes the particles stay together and it is very difficult to separate them. Many solvents or mixtures of solvents for polymers can be considered at a given temperature [19]. The solvent is evaporated and condensed in order to use it again, while in a mold the nanocomposite piece is obtained. This method is more appropriate for armored nanocomposites, because polymers such as polyaramides, ultrahigh molecular weight polyethylene and polycarbonate can be dissolved to mix with armor fillers. The process is shown schematically in Figure 4.12.

In-Situ Polymerization — In this method the CNTs are added at the beginning of the polymerization reaction directly to the reactor and mixed with

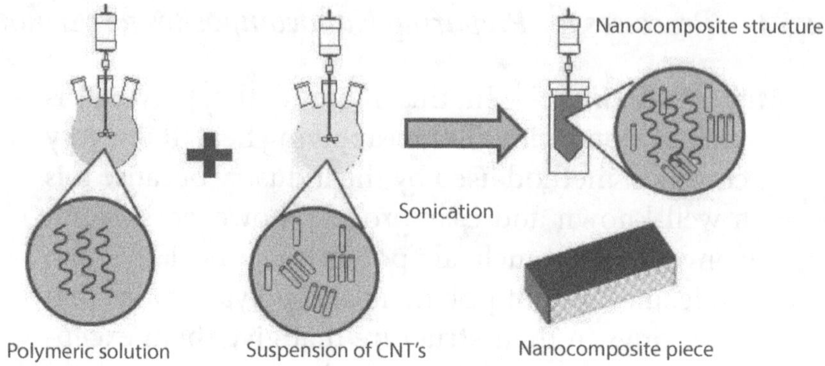

Figure 4.12 Solution mixing process to obtain CNTs/polymer armor nanocomposites.

reagents, even before obtaining the monomer, as is the case with polyesters. During the chemical reaction, stirring is applied for hours; intensive mixing causes agglomerations of CNTs, that needs to be dispersed uniformly throughout the polymer matrix. In these reactions, temperature, vigorous agitation and reaction times of 8 hours are applied; the combination of these factors causes a good interface interaction of the CNTs and the polymeric matrix, providing better mechanical properties that can be used to obtain armored nanocomposites. Also, other properties are substantially improved with a good interface of both materials. Figure 4.13 shows the equipment where the *in-situ* polymerization reaction occurs with the filler, exemplified for polyesters. The extrusion of fibers can be done directly from the reactor through the application of nitrogen pressure.

Another advantage of *in-situ* polymerization is that bonds between filler and polymer can occur; an example of this is reported for Kevlar and multi-walled carbon nanotubes [20]. This process

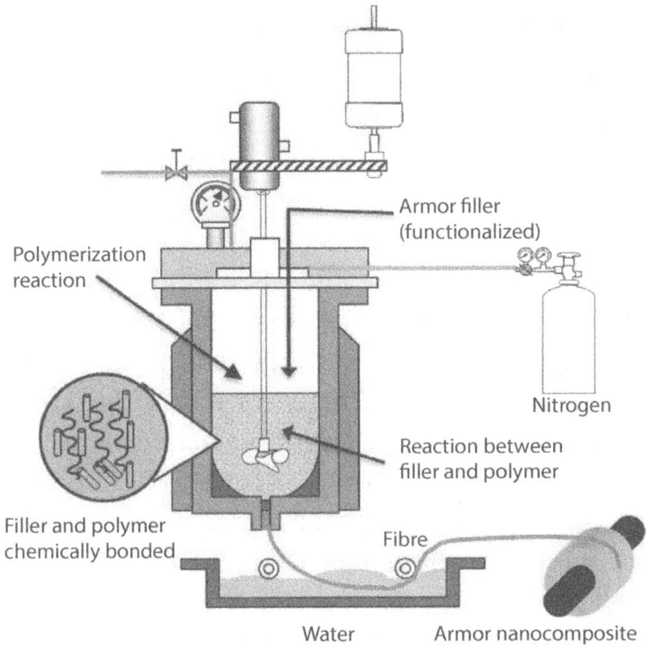

Figure 4.13 Armor nanocomposites obtained via *in-situ* polymerization.

is similar to the previous one. It is known as polycondensation in solution because hexamethylenephosphoramide is added to the reactor to form an initial solution, in both cases a chemical reaction occurs to form chemical bonds.

Other Methods for Obtaining Nanocomposites — One of the methods is to use latex, which is a mixture of polymer nanoparticles of different sizes inside of a liquid. A sonicated suspension of armor nanofiller particles is directly poured into the latex and the mixture is mechanically stirred. Finally, the latex is coagulated, modifying the pH with acids or additives, and the liquid is separated from the nanocomposite. This area is still under development because the above methods can be used in combination and also with new procedures.

4.2.4 Armor of Smart Nanomaterials

Smart materials for armor are multifunctional nanomaterials for the protection of life that can measure and process information to act autonomously before a stimulus and can return to their original state or to another one required [21]. If it were necessary to protect a nanomechanism used for military purposes, then the definition would also apply to the security of devices and not just of military personnel [22]. In the military area they are called reactive materials, because specifically the armored bulletproof material reacts or responds to the presence of the penetrating projectile. The other materials are called passives and only work to stop the projectile without a sensor that starts with the information process and subsequent intelligent action. An intelligent fiber has a system that includes sensor, information processing and actuator (Figure 4.14). This system is improved by use of information storage and communication toward a control center.

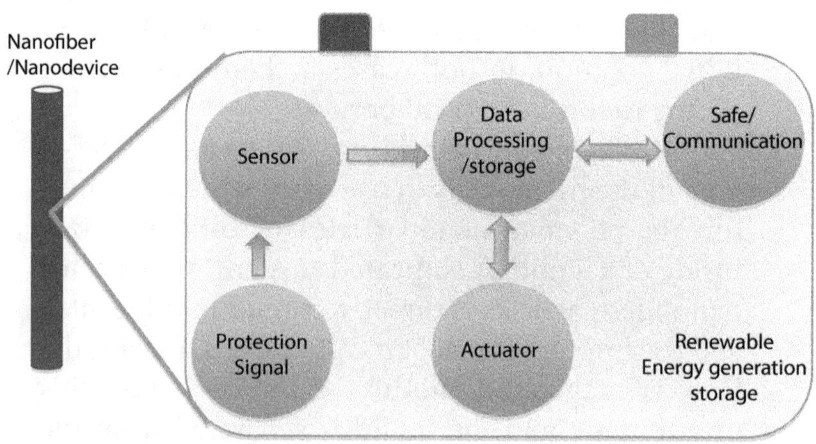

Figure 4.14 Typical smart nanomaterials for armor.

4.2.4.1 *Memory Materials*

Some materials, such as metals, alloys, ceramics, polymers, plastics and composites, exhibit a property called memory, which was discovered by Chang and Read [23]. Shape memory refers to the recovery of the original shape of a piece after it has been deformed. Thus, the materials with memory of form that depend on the temperature can easily be deformed at a given temperature, but if it returns to its original temperature, the piece will recover its original form (designer materials). Nickel-titanium alloy has good mechanical properties of processability and resistance to fatigue and corrosion, making it a multifunctional material. Ceramic materials have the disadvantage of being fragile and have little recovery in their original form because microcracks occur during the change of shape.

In the case of polymers, the phenomenon of cracking is less, which makes them attractive for the study of this phenomenon. Some examples are the materials used in catheters inserted into parts of the body such as the urethra (Figure 4.15). Other materials based on polyurethanes have excellent memory properties and also present good results in cytotoxicity tests when used for medical applications.

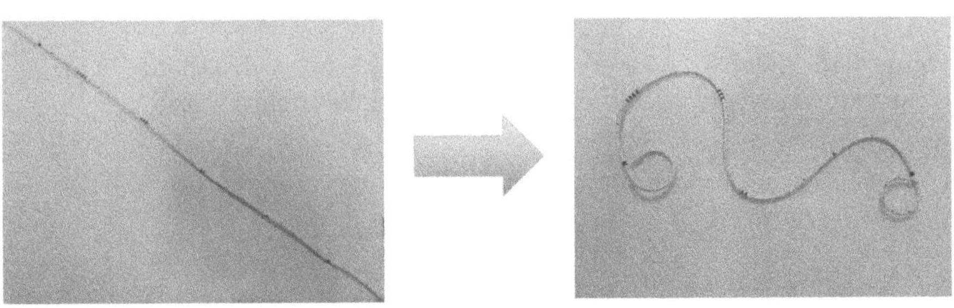

Figure 4.15 Straight polymer changes its shape with body temperature (shape memory).

These nanomaterials with some crosslinked polymers can also change their shape using a different variable such as pH-induced shape memory (Figure 4.16).

An application of materials with very attractive memory in electromechanical mechanisms, shown schematically in Figure 4.17, are known as actuators. Electroactive, electrostrictive or piezoelectric materials are used for their construction. In electroactive materials an electric field is applied causing the anion and cations of a polymer to be expelled or absorbed, thus a shrinking or swelling of the material can occur. In the second case, the electroactive material is a dielectric piece producing surfaces which are attracted to each other by the application of an electric field. In the case of piezoelectric materials, they can provide electricity by applying external forces, which is a fundamental reason why they can be used in nanomaterials for the protection of life when incorporated into textile fibers, so they

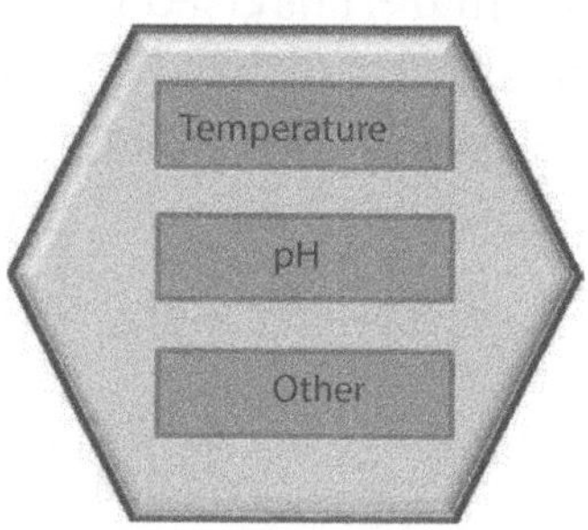

Figure 4.16 The shape memory of polymers can change with a change in pH, temperature or other variables and then recover their original shape.

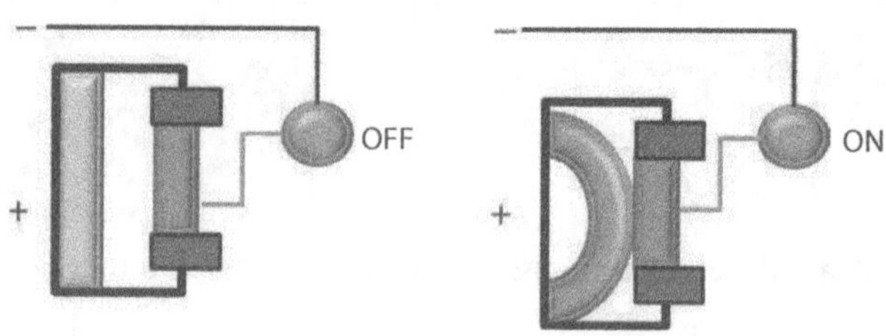

Figure 4.17 Operation of an actuator showing on and off positions.

can add another function. This could provide power to tactical equipment with low power consumption.

The reaction of intelligent nanomaterials can also be a change of color and modification of the chemical or electronic structure, which can involve processing, receiving and sending information.

4.3 Summary

Requirements for armor for protection of military personnel is very imperative. Making the material stronger and smarter are basic requirements of armor in the present-day scenario. Hence, in this chapter nano-sized materials and their composites that can be incorporated into smart armor are touched upon. Polymers and carbon nanomaterials have found a very special space in fabricating nanotechnology-based armor.

References

1. Ashcroft, J., Daniels, D., Hart, S., *Selection and Application Guide to Personal Body Armor*, pp. 1–20, NIJ Guide 100–01, National Institute of Justice, Rockville, MD, USA, 2001.
2. Ostafin, A., Landfester, K., Sridar, L., Moukasian, A. (Eds.), *Nanoreactor Engineering for Life Sciences and Medicine*, pp. 5–6, Artech House, Boston, London, USA, SBN13 9781596931589, 2009
3. Ma, J., Zhang, L., Geng, B., Azhar, U., Xu, A., Zhang, S., *Molecules*, 22, 152, 152–165, 2017.
4. Comellas-Aragones, M., Engelkamp H., Claessen V.T., *Nat. Nanotechnol.*, 2, 10, 635–639, 2007, doi: 10.1038/nnano.2007.299.
5. Karlsson, M., Davidson, M., Karlsson, R., Karlsson, A., Bergenholtz, J., Voinova, M., Orwar, O., *Annu. Rev. Phys. Chem.*, 55, 613–649, 2004.

6. Loiseau, A., Launois-Bernede, P., Petit, P., Roche, S., Salvetat, J.-P., *Understanding Carbon Nanotubes –From Basics to Applications*, p. 552, Secaucus, New Jersey, U.S.A.: Springer Verlag, 2006.
7. Hearle, J.W.S., *High-Performance Fibres*, vol. 327, CRC Press, Boca Raton, Boston, New York, Washington DC, 2001.
8. Zhu, J., Ding, Y., Agarwal, S., Greiner, A., Zhang, H., Hou, H., *Nanoscale*, 9, 18169–18174, 2017.
9. Li, K., Wang, Y., Xie, G., Kang, J., He, H., Wang, K., Liu, Y., *J. Appl. Polym. Sci.*, 135, 46130, 2018.
10. Iijima, S. and Ichihashi, T., *Nature*, 363, 603–605, 1993.
11. Hata, K., Futaba, D., Mizuno, K., Namai, T., Yumura, M., Iijima, S., *Science*, 306, 1362–1364, 2004.
12. Khan, W., Sharma, R., Saini, P., Carbon nanotube-based polymer composites: Synthesis, properties and applications. IntechOpen, London, UK, 2016, doi: 10.5772/62497.
13. Sharon, M. and Sharon, M., *Def. Sci. J.*, 58, 4, 5491–5516, 2008.
14. Hazell, P., *Armour: Materials, Theory, and Design*, pp. 54–62, CRC Press, Boca Raton London New York, 2015.
15. Nataraj, S.K., Yang, K.S., Aminabhavi, T.M., *Prog. Polym. Sci.*, 37, 487–513, 2012.
16. Kumar, A., Chavan, V., Ahmad, S., Alagirusamy, R., Bhatnagar, N., *Int. J. Impact Eng.*, 89, 1–13, 2016.
17. Mylvaganam, K. and Zhang, L.C., *Nanotechnology*, 18, 1–4, 2007.
18. Kimura, Y., Tsuchida, A., Katsuraya, K., *High-Performance and Specialty Fibers*, p. 451, Springer, Japan, 2016.
19. Mark, J.E., *Polymer Data Handbook*, p. 1012, American Chemical Society, J. Am. Chem. Soc. 131, 44m 16330, 2009.
20. Sainsbury, T., Erickson, K., Okawa, D., Sebastian, C., Frechet, J., Zettl, A., *Chem. Mater.*, 22, 2164–2171, 2010.
21. Aliberti, K. and Bruen, T., Designer Materials. *Army Logistician*, 38, 36–41, 2006.
22. Kiekens, P. and Jayaraman, S., *Intelligent Textiles and Clothing for Ballistic and NBC Protection*, NATO Science for Peace and Security Series: B, Physics and Biophysics, pp. 119–136, Springer, Dordrecht, The Netherlands, 2012.
23. Chang, L.C. and Read, T.A., *Trans. AIME*, 189, 47–52, 1951.

5

Nanotechnology and Weapons

Chetna Sharon

Division of Hematology, Oncology & Palliative Care, Virginia Commonwealth University and Department of Internal Medicine, Hunter Holmes McGuire VA Medical Centre, Richmond, Virginia

In thinking about nanotechnology today, what's most important is understanding where it leads, what nanotechnology will look like after we reach the assembler breakthrough.
K. Eric Drexler

5.1 Introduction

The first basic needs of primitive man for survival were quenching hunger with food from plants and using weapons to kill animals for food. Next came requirements like shelter, source of energy (fire) and clothes; and as their intellectual capacities grew, vehicles, furniture and so on. This mysterious and curious creature started exploiting his gray matter to make weapons from stone, wood and metal, until reaching the use of explosive chemicals and nuclear energy of today. The most recent and advanced field of science—Nanotechnology—is now attracting them. Scientists have realized the perfection of nano-sized particles manifested in their mega impact and immense application. No

Madhuri Sharon, et al. Nanotechnology in the Defense Industry: Advances, Innovation, and Practical Applications, (133–164) © 2019 Scrivener Publishing LLC

wonder nanotechnology, which explicitly deals with unique all-pervading phenomena, has become a matter of excitement and interest for use in future generations of weapons, culminating in the monstrous entry of nanotechnology in modern warfare.

It will not be an exaggeration to say that Eric Drexler [1] was a pioneer who established a foundation for the possibility of constructing molecular assemblers and nanometer-sized machines, which gave rise to the possibility of fabrication of many nano-sized entities that can find application in various fields, including weapons. This possibility gave impetus to many like that of nanoweaponry, which will make all conventional weapons of war obsolete and *future wars will be waged with nanoweapons* [2]. Now it is believed that nanotechnology has the potential to eventually replace conventional nuclear weapons with nanoweapons because they are easy to build, easy to deliver but hard to monitor and become obsolete almost immediately.

The advantages that defense personnel are seeing in the use of nanotechnology for making WMDs (weapons of mass destruction) are that it will reduce the amount of payload on the stealth fighter planes carrying them and delivery will be more accurate.

5.2 Considerations for Developing Nano High Energy Materials (HEMs) for Weapons

Nanomaterials have been envisaged to be incorporated into HEMs, i.e., materials which have a high energy release rate when initiated, like propellants, explosives and pyrotechnics. The unique properties of nanoparticles, especially high surface area and enhanced surface reactivity, are the main reason for the use of nanoparticles in HEMs, e.g., in propellant it offers improved burning rate; in explosives it improves blast properties and mechanical strength formulations. The work being done to make nano-ingredients includes AP (ammonium perchlorate), Al (aluminum), RDX (cyclotrimethylene trinitramine), HMX (cyclotetratrimethylene trinitramine), CL-20 (hexanitrohexaazaisowurtzitane), HNF (hydrazinium nitroformate), metal oxides, etc. The requirements for the performance properties of HEMs are specific impulse, burning rate, force constant, velocity of detonation, density, luminosity, etc. A single nano-ingredient may not fulfill all these requirements. Moreover, requirements for a propellant or explosive and pyrotechnic will be different. Therefore, a composite, conjugate or mixture of different materials are required to make HEM formulations. The materials that are loaded in propellants, explosives and pyrotechnics are solids in powder form. The particle size of different materials used is different.

One of the major impacts of using nanoparticles in the area of defense is miniaturization of product that has novel properties such as high chemical reactivity, large surface to volume ration and unique combustion characteristics. Apart from HEM ingredient, there are requirements for HTPB (hydroxyl terminated polybutadiene) binders, TDI (toluene diisocyanate) curing agents, and DOA (dioctyl adipate) plasticizers. Other formulations used in double-base propellants are nitrocellulose, nitroglycerine, carbamites, etc.

To fabricate nanoweapons, reduction in defect concentration of nanomaterials is being exploited for HEMs.

5.2.1 Propellants

Conventional Propellant is a double-base propellant composed of nitrocellulose and nitroglycerine. Propellants are composed of binders in which oxidizers or metallic fuels are loaded.

Oxidizers — The most commonly used oxidizers are (i) ***Perchlorates:*** Potassium, Ammonium and Nitronium; (ii) ***Chlorates:*** Sodium Potassium and Barium; (iii) ***Nitrates:*** Potassium, Ammonium, Sodium, Calcium, Strontium, Cesium and Barium; (iv) ***Permanganates:*** Potassium and Ammonium;

(v) ***Dichromates:*** Barium, Lead and Potassium; (vi) ***Oxides and Peroxides:*** Barium peroxide, Strontium peroxide, Lead tetroxide, Lead dioxide, Bismuth trioxide, Iron(III) oxide, Iron(II,III) oxide, Manganese(IV) oxide, Chromium(III) oxide and Tin(IV) oxide; (vii) ***Sulfates:*** Barium sulfate, Calcium sulfate, Sodium sulfate and Strontium sulfate (sulfates are used at very high temperatures); (viii) ***Organic Chemicals:*** Guanidine nitrate, Hexanitroethane, Cyclotrimethylene trinitramine and Cyclotetramethylene tetranitramine; (ix) ***Others:*** Sulfur, Teflon and Boron.

Since they have much less or insignificant amounts of solid particles, nanoparticles could not contribute to its improvement. Whereas, composite propellant has

(i) 80–85% solid (oxidizers, solid fuel as metal and transition metal oxides as catalyst) for burning rate modifier. The most common oxidizer used in rocket propellant is ammonium perchlorate because it has high oxygen content, desired heat of formation, and high stability in the working pressure and temperatures of missiles and rockets. Some other energetic solid oxidizers being tried are ADN (ammonium dinitramide) and hydrazinium nitroformate (HNF); the exhaust from these two oxidizers is more eco-friendly. But the problem with ADN is that it is very hygroscopic and not compatible with butadiene-based composite propellant binder such as hydroxyl-terminated polybutadiene (HTPB).

(ii) Aluminum perchlorate (AP) and aluminum (Al) as metallic fuel that increases the burning rate. Metallic fuels like Al are added into the propellant to enhance density, increase energy content and suppress instability.

In nano-sized crystalline $CuCr_2O_4$ (copper chromate), thermal decomposition of AP is enhanced by nano-$CuCr_2O_4$, [3] and nanopowder of Ni, Fe and Cu [4]. CuO (cupric oxide) is more effective than micro size in increasing both exothermic peak and reduction of high temperature exotherm.

(iii) Cu/CNT composite prepared by precipitation method exhibits very high catalytic performance on the decomposition of AP. For propellant, nano-Al synthesized by electro-condensation produces spherical 43 nm nanoparticles with passivating oxide layer of 3 nm with its surface doped with either barium, benzene, silicon or rubber, which was found to be more efficient in burning rate and it also decreased agglomeration [5].

Catalysts — Some oxidizers used in propellant formula act as catalyst also, e.g., ammonium dichromate is used as a catalyst in ammonium nitrate-based propellant formulas. However, generally propellant formulas require transition metal or their complexes as catalyst to burn faster and more stably, e.g., iron(III) oxide, hydrated ferric oxide, manganese dioxide, potassium dichromate, copper chromite, lead salicylate, lead stearate, lead 2-ethylhexoate, copper salicylate, copper stearate, lithium fluoride, n-butyl ferrocene, and di-n-butyl ferrocene.

5.2.2 Explosives

There have been trials and efforts in incorporating nanosized particles in explosives. It has been found that Nano RDX is less sensitive for impact and shock than micro RDX. Some of the findings are that Fe nanopowder reduced the decomposition temperature of HMX by 90 °C. Al, Ni, Cu and Wo nanopowder has also been found to reduce the onset of HMX [4]. To enhance the blast and burning properties of explosives the surface property of PETN (pentaerythrite tetranitrate) needs to be altered. Surface coating of nano-HNS (hexanitrostilbene) is done by CMC-Na (carbonyl methyl cellulose) and white dextrin; this has also increased the short duration of pulse sensitivity of HNS [6].

In explosives, the metallic fuel Al, which has high enthalpy of combustion, is added to enhance blast performance, improve underwater performance, enhance temperature of detonation product, alter equilibrium product distribution and influence energy release rate. Al has been found to be better than other metallic fuels used in explosives, such as beryllium, boron, tungsten, etc.; however, according to Jones *et al.* [7] the nano-sized Al, due to the presence of passivating Al oxide, has shown lower performance than expected. When nano-sized Al is used, depending on its size and thickness of passivation layer, it affects the onset of temperature, enthalpy charge and mass grain on oxidation [8]. The 90 nm Al nanoparticle was more reactive than the 180 nm one. Moreover, nano-sized Al increased the rate of thermal decomposition of TNT and RDX and onset of lowered temperature. Electrostatic sensitivity of HEM, TNT, HMX and RDX increased with addition of Al nanopowder. When RDX, HMX and PETN were doped with Cr_2O_3, they showed increased reactive properties [9].

Other nano-explosive composites, like nanoparticles of CL-20 uniformly coated HDI crosslinked nitrocellulose of 20–300 nm and 90% solid loading, were synthesized by Tappan and Brill [10]. Similarly, HEMs were also coated with advanced binders like poly-AMMO-BAMO (3-azidomethyl-methyl oxetane-poly(3-azidomethyl-3-methyloxetane) and poly-NMMO-GAP (3-nitratomethyl-3-methyloxetane-glycidylazide polymer). These nanocomposites gave high heat of explosion [11].

Kanel *et al.* [12] have shown that if micro-sized B in the range of 0.1–1 μ is incorporated into HMX, it has a negative effect on the detonation parameters and heat addition due to combustion. But nano-sized B or Al work better by affecting the burning rate and thermal decomposition of other materials.

Additives — According to Stamatis *et al.* [13], nanopowder additives like 8Al + MoO_2 and 2B + Ti in Al increases burning rate and rate of pressure rise. Mechanical and reactive properties of many nanocomposites of Al and B are also being tested such as 2Al + MoO_{3+}, 2Al + Fe_2O_3, 2Al + CuO, 8Al + WO_3 and 2B+Ti as well as 2B+ Zr.

Other commonly used additives are added such as (i) ***Coolants*** such as clay, diatomaceous earth, alumina, silica, magnesium oxide, or endothermically decomposing materials, such as carbonates, are added; Oxamide is used as a high performance burning rate suppressant and strontium carbonate is used as a fire retardant in some gunpowders; (ii) ***Flame suppressants*** such as potassium nitrate and potassium sulfate; (iii) ***Opacifiers*** such as carbon black and graphite are often used; (iv) ***Colorants***, usually salts of barium, strontium, calcium, sodium, copper metals are used as sources of chlorine, which simultaneously serve as an oxidizer; (v) ***Stabilizers*** such as chlorates, carbonates (of sodium, calcium, or barium carbonate), boric acid, organic nitrated amines, 2-nitrodiphenylamine, petroleum jelly, castor oil, linseed oil, ethyl centralite and 2-nitrodiphenylamine are used.

Apart from these, ***Anticaking agents*** (fumed silica); ***Binders*** (gum arabic, red gum, guar gum, copal, carboxymethyl cellulose, nitrocellulose, rice starch, cornstarch, shellac, dextrin); binders can also serve as fuels; ***Plasticizers*** such as camphor can be used; ***Binders, curing bonding agents and crosslinking agents*** are used in the manufacture of compact compositions of pyrotechnics.

5.2.3 Pyrotechnics

Pyrotechnics is the use of combinations of materials that can generate self-contained and self-sustained exothermic chemical reactions that produce heat, light, gas, smoke and/or sound. It is commonly used in many consumer products such as model rocket motors, highway and marine distress flares, sparklers and caps for toy guns, safety matches, oxygen candles, explosive bolts and fasteners, fireworks, components of automotive airbags, and gas pressure blasting in mining, quarrying and demolition. For these products, basic pyrotechnic compositions are: flash powder, gunpowder, gas generators, e.g., chemical oxygen generators, ejection charges, burst charges, smoke compositions, delay compositions, pyrotechnic heat sources, white or colored sparklers and flares.

Materials used are composed of small particles of

(i) ***Metals*** such as aluminum, magnesium, iron, alloy of iron (steel), zirconium, titanium, ferrotitanium (iron-silicon alloy), ferrosilicon, manganese, zinc copper, brass, tungsten, nickel-zirconium alloy;

(ii) ***Metal-Hydrides*** such as titanium(II) hydride, zirconium(II) hydride, aluminum hydride, decaborane, etc.;

(iii) ***Metal-Carbides*** such as zirconium carbide;

(iv) ***Metalloids*** such as silicon, boron, antimony, sulfur, red phosphorus, white phosphorus, calcium silicide, antimony trisulfide, arsenic sulfide, phosphorus trisulfide, calcium phosphide and potassium thiocyanate;

(v) ***Carbon Based Fuels*** such as carbon, charcoal, graphite, carbon black, asphaltum and wood flour [14];

(vi) ***Organic Chemicals*** such as sodium benzoate, sodium salicylate, gallic acid, potassium picrate, terephthalic acid, hexamine, anthracene, naphthalene, lactose, sucrose, sorbitol, dextrin, stearin, stearic acid and hexachloroethane;

(vii) ***Organic Polymers and Resins*** such as Teflon, Viton and other fluoropolymers, hydroxyl-terminated polybutadiene, carboxyl-terminated polybutadiene (CTPB), Polybutadiene acrylonitrile (PBAN), polysulfide, polyurethane, polyisobutylene

nitrocellulose, polyethylene, polyvinyl chloride, polyvinylidene chloride, shellac, and accroides resin (red gum).

For nanoweapons, nanoparticles are used for pyrotechnics as (i) Initiator, (ii) Obstructant and (iii) Thermite.

(i) ***Initiator*** — It is a pyrotechnic device which causes first fire in the ammunition containing propellant or explosive. Metallic nanolaminate foil is used as initiator. It is coated with a nanocomposite synthesized by sol-gel method. This sol-gel energetic composition is coated on Monel-400/A-1 nanolaminate substrate of 25 μ thickness. Here Fe_2O_3 (iron oxide) is used as sol in which Al is added before dipping nanolaminate strip [15].

(ii) ***Obstructant*** — It is a pyrotechnic venture which camouflages products on the battlefield. Obstruction material should have a desired extinction cross section that interacts with radiation.

Appleyerol and Nigel [16] did the modeling of IR extinction of highly conducting micro- and nanofibers and disk flake. They observed that extinction cross section of fiber was lower than the disk flake of equivalent low frequency electrical conductivity and was a major dimension of the particle and aspect ratio. Reduction in fiber diameter further increased the volume extinction. They suggested that optimum IR extinction performance can be obtained when fiber has high conductivity and has length equivalent to the wavelength of the IR radiation.

(iii) ***Thermites*** — Nano-thermites are high-tech energetic materials, which facilitate effective propellants and highly explosive load of ammunition

and projectiles. Nano-thermites can be ignited in many ways. Not only are they "high explosives," but they are pyrotechnic materials as well. To get high porosity nano-thermites, silica is used. Other combination materials used include RDX with thermoplastic elastomers and nano-aluminum (oxidizer) combined with Al/Fe_2O_3, Al/CuO, Al/MoO_3, Al/WO_3 and Al/Bi_2O_3. $KMnO_4$ nanoparticles are also used as oxidizer to prepare Al/KMnO4. Thermite-type highly exothermic reactions are used for welding, melting solid and other heat-reactive reactions. Iron(II,III) oxide is used as an oxidizer in thermite and thermate.

5.3 Requirements for Nanoparticles Used in Nanoweapons

Nanotechnology is being tried for making more lethal as well as non-lethal weapons. Fabrication of lethal weapons demands precision targeting, reduced weight and signature, enhanced impact damage and cost effectiveness; whereas, non-lethal weapons are being manufactured to neutralize the enemy temporarily. For both types of weapons, there is a need to have supporting nano-systems of materials having the following attributes:

- High strength;
- Light weight;
- Able to absorb radar signature;
- Super penetrating capacity;
- Weapon carrying system with built-in firing monitoring sensors;
- Nanomaterials should sharpen upon impact or provide additional damage high energetic propellants; for this purpose, aluminum nanoparticles

are used as propellant and quantum structures are used for directed energy weapons, e.g., microwave and high energy directed laser systems;

- Should be anti-kickback systems;
- Better targeting using sensors: μ-radar, μ-bolometers (IR) and acoustic arrays are used;
- Remote and unmanned guidance sensor and wireless communication are other support requirements; and
- Remote-controlled robotic systems with firing capability.

Nanotechnology is important for reducing the size and weight of high impact weapon systems, nonlethal high energy μ-lasers, microwave, RF or acoustic wave generator weapons to neutralize target groups for a certain amount of time. Nanotechnology is being used for high impact ammunition and for developing high energetic nanocomposite materials.

Unmanned weaponized robots (TALON™ robots) are powerful, durable, lightweight, remotely operated, tracked vehicles. An operator positioned a kilometer away from the danger point can operate them. TALON robots are used during day and night in all weather and in all types of terrain for reconnaissance, communications, sensing, security, defense and rescue operations. According to the Army News Service, *TALON robots* have been in continuous, active *military* service since the year 2000, when they were successfully used in Bosnia for the safe movement and disposal of live grenades.

5.4 Synthesis of Nanomaterials for Weapons

Generally, the synthesis of nanoparticles for HEM application use μ-size particles of their counterparts by grinding, wet-jet milling, crystallization, spray freezing in nitrogen, spray drying, sol-gel and supercritical fluid methods. The physical methods mentioned above, such as grinding, create challenges in controlling particle size, particle distribution and safety during bulk manufacturing. Whereas challenges of the crystallization method are slow nucleation and growth of stress-free crystals; however, structure and shape control are easier by crystallization. There are various methods of synthesis of different explosives and propellants [17], some of which are mentioned below.

Supercritical Fluid Method — This method is used to synthesize nano-sized RDX (cyclotrimethylene trinitramine).

Use *of Surfactant for Nano-RDX Manufacturing* — This method was attempted by Luo *et al.* [18]. They used bacterial cellulose, gelatin as matrix and acetone as solvent. To this combination, surfactants were added to enhance its required properties. The surfactants used were SDBS (sodium dodecyl benzene sulfonate) and OP (p-octyl polyethylene glycol phenylether). Addition of surfactant increased the nano RDX (30–60 nm) content in the prepared film from 75% to 84%.

Spray Freezing into Liquid (SFL) — This method has been used to synthesize 70 m to 90 nm of the explosive NTO (%-nitro-2,4-dihydro-3H-1,2,4-triazole-3-one) by spraying NTO into liquid nitrogen [19].

Precipitation Method — Reticularly structured HMX (cyclotetratrimethylene tertratrinitramine) propellant was synthesized at room temperature by a simple precipitation method by Yongxu *et al.* [20] using HMX dissolved in acetone and then dropped in organic non-solvent to precipitate at room temperature; this yielded ~50 nm nanoparticles. But these nanoparticles were less sensitive to impact than conventional HMX. This indicated that all the nanoparticles may not give desired increased efficiency. Later, Bayat *et al.* [21] controlled the particle size of submicron HMX dissolved in acetone by precipitating it in non-solvent cold hexane. The precipitate was filtered and dried. Crystals thus obtained were 300 nm wide and 1–2 μ long.

Crystallization Method — This method is used for manufacturing many HEMs, e.g., TATB (1,3,5-triamine-2,4,6-trinitrobenzene) crystals were prepared by solvent/non-solvent crystallization with concentrated sulfuric acid, which yielded 27–41 nm nanoparticles [19].

Dip Pen Lithography — This technique has been tried for manufacturing HMX and another HEM, PETN (pentaerythrytol tetranitrate).

Spin Coating Method — This method has been used for coating nano-sized continuous film of PETN or HMX dissolved in acetone on cleaved mica substrate; thus, PETN or HMX was patterned on the mica. HMX was also patterned on silicon by this method [22].

Spraying Method — This method was found to be suitable for synthesizing K-6 (2-oxo-1,3,5-dinitro-1,3,5-triazacyclo-hexane) or Keto-RDX, a HEM having high density and suitable heat formation. K-6 dissolved in acetone was sprayed on stirred water at room temperature and this yielded nano-sized K-6, resulting in the high surface quality of nano-K-6 becoming less sensitive.

Sol-Gel Method — This method is used to synthesize nano-thermites. Nano-thermites are made by mixing fine-grained Al and metal oxides (iron oxide, molybdenum oxide or copper oxide, etc.). The mixing is done by sol-gel method. First the reactants are added to a liquid solution to get "sols" and then a gelling agent is added to these sols, resulting in "sol-gel," which is dried to form a nanoporous reactive material having high surface area that offers higher rate of energy release than is seen in macro-thermite mixtures. Because of these properties, sol-gel nano-thermites are also known as super-thermite, energetic nanocomposites and metastable intermolecular composites. Sol-gel method is also used for preparation of RDX. Moreover, bacterial cellulose with gelatin as matrix and acetone as solvent have been used to synthesize RDX of 30–40 nm [23].

During synthesis, mostly Al-oxides as well as other metal oxides, such as Fe, Mn, Cu, Mo, bismuth and tungsten, are also used as oxidizer. Al oxides are also used as fuel. Oxidizers are used for decreasing the particle size and increasing the particle contact as well as combustion velocity.

5.5 Characterization of Nanomaterials Used in Weapons

Nano-sized materials need a lot of consideration as far as their explosive or propellant properties are concerned, e.g., nano-K-6 explosive. Though it is believed that nano-sized materials are more reactive due to their SVR, the high surface quality of nano-K-6 became less sensitive than its micro counterpart. Similarly, the X-ray diffraction angle is the same for both nano- and micro-sized particles, but with a decrease in nanoparticle size the strength of peak decreases along with the friction sensitivity. Thermogravimetric analysis (TGA) and differential scanning calorimetry (DSC) were also the same for both, but thermal stability of nano-K-6 was lower [24]. Scanning electron microscopy (SEM), field emission scanning electron microscopy (FESEM), transmission electron microscopy (TEM), atomic force microscopy (AFM), X-ray diffraction (XRD) and small angle scanning (SAS) were invariably used by Mang and Hjelm [25]. For morphological analysis of particle size, shape and distribution, SEM and TEM was used. Recrystallized particles and melting point were analyzed using FESEM, XRD and high-performance liquid chromatography (HPLC). AFM augmented with thermal cantilever helps in thermomechanical data reading and writing and controls bulk material properties like toughness, Young's modulus of elasticity, and tear resistance at nanoscale. The Brunauer-Emmett-Teller (BET) method is the most commonly used method for surface area measurement. SAS has been used for distinguishing between aggregate and primary particle, to differentiate between the morphology of primary particles and particle size distribution. This equipment also gives insight into deflagration of explosives, solid-solid phase transition of explosives and nanoscale behavior of shocked explosives. For measuring particle chemistry, a helium pycnometer is used.

5.6 Nanomaterials for Use in Nanoweapons and Ammunition

The properties of nanomaterials that are being envisaged for use in creating more efficient and improved ammunition and projectiles are their penetration capacity, reducing weight and enhanced explosive force. Moreover, nanotechnology has been used for fabricating various built-in sensor and power supply for sensors.

5.6.1 Super Penetrating Materials

Though depleted uranium (DU; also referred to as U-235) based alloys are being used for armor-penetrating projectiles, since it is slightly radioactive and toxic material, in high doses it can cause adverse health effects. Therefore, nanomaterials are being developed to replace DU, e.g., nanocrystalline tungsten.

5.6.2 Nanocrystalline Tungsten

A metallic glass matrix reinforced with nanocrystalline tungsten works even better at nanometer scales than depleted uranium (Ames Laboratory); this composite is environment friendly and not hazardous to health. Other alternatives to depleted uranium could be nanocrystalline tantalum that has high density and ductility, making it an attractive material for ballistic applications. In nanocrystalline state, tantalum does not encounter strain hardening; hence, the dynamic formability and processability is easy [26].

5.6.3 Liquid Metal

Liquid metals are alloys of metals having very low melting points; hence they are in liquid state at room temperature. Examples of liquid metals are mercury, gallium, bromine, cesium, and rubidium. Liquid, amorphous metal alloys, such as Zr-base and Ti-base liquid metal alloys, show very high yield strengths, which are more than twice the strength of conventional titanium alloys. Liquid metals combined with nanocrystalline particles form high penetrating materials. These materials have high hardness and yield strength, superior strength/weight ratio, higher elastic limit, high corrosion and wear resistance and unique acoustical properties.

5.6.4 High Energy Laser Weapons

High energy laser weapons carried by trucks are expected to be on the battlefield very soon [27]. A laser is being employed by

Boeing for building the optics needed to track and focus lethal laser energy onto rockets, artillery shells and mortar rounds. The laser is already used for shooting at real targets.

Quantum dots are expected to be highly efficient light and laser sources and detectors for telecom applications.

5.7 Nanoweapons

Bonadies [28] has defined nanoweapons as such: "Nanoweapons may be constructed applying a combination of techniques and materials to produce very small (a nanometer is one-billionth or 10-9 of a meter) mechanical devices or inorganic nano-materials or bio-materials that could interact with, and potentially damage or otherwise alter, a number of different biological or non-biological systems" [28].

Although this definition of nanoweaponry may now appear to be only possible in science fiction or fantasy, at the rapid rate at which nanotechnology is progressing, unfortunately, we may see it become a reality in the very near future. The idea of using teleweapons supported by expanding sensor capabilities and wireless communication is also in the offing. It will enable the remote operation of weapons using feedback from previous impacts and can be self-adaptive targeting.

It has been found that nanometals, such as nano-aluminum, can be used for making compact powerful bombs that have ultrahigh chemical explosive which is more powerful than conventional bombs. However, in the current scenario, nanoweapons are still in the R&D stage. Currently under development are, for example, nano-engineered high explosives for advanced weaponry using nano-engineered components of extreme precision. Nanotechnology can satisfy the growing demand for better materials with extremely well-characterized specifications. Nanomaterials and polymer conjugate are being envisaged to create

(i) Lightweight guns, rifles and automatic firing systems. High power laser for handguns is also being thought of;

(ii) RFID tags for guns and cartridges target positioning/recognition via micro IR cameras on guns **Personal Digital Assistants (PDAs)**.

(iii) Target positioning/recognition via micro-radar, RF-array, and through-the-wall THz radar. The gun will be equipped with a non-mechanical trigger which can be wirelessly linked to the PDA, phone or smart helmet of a soldier. An example of the use of present wireless technology to detonate explosives is the use of GSM (global system for mobile communications) technology has been used by terrorists in Iraq.

(iv) Teleweapons like a wireless link to a PDA and smart helmet to pull the trigger of a gun from a distance. The gun will be equipped with a non-mechanical trigger which can be wirelessly linked to the PDA, phone or smart helmet of the soldier.

(v) Various types of ammo, such as shaped ceramic materials, softer bullets, high penetration bullets, sensor modes/smart dust, and tailored insensitive explosives, limit collateral damage.

5.7.1 Types of Nanoweapons

The types of nanoweapons that are envisioned or even being tried can be categorized as (i) Biological, (ii) Chemical (iii), Mechanical, and (iv) Molecularly assembled weapons. Biological and chemical weapons are discussed in the following chapter. Here we are concentrating on mechanical and molecularly assembled weapons. Nano-built weapons are easy to fabricate.

Once a nanoweapon is developed, one of its major roles will be to facilitate and enhance the delivery mechanism. The modes of delivery will need both space-based and aero delivery as well as crawlers. Thus, the delivery system will require considerations like payloads, methods of targeting, modes of delivery and means of production.

5.7.1.1 *Molecularly Assembled Nanoweapons*

Molecular nanosystems as foreseen by Roco [29, 30] involve "multifunctional molecules, catalysts for synthesis and controlling of engineered nanostructures, subcellular interventions, and biomimetic for complex system dynamics and control." This can have many capacities, such as mechanical operation, independent power generation, information processing and communication, even applicability can be immense at the living system level; for example, it can transform hazardous materials and mix a specified amount of oxygen into the soil, nanodevices injected into the body can fix the DNA of damaged cells, monitoring vital conditions and transport messages to computers that might operate by reading the brain waves of the operator. This will need the design of intelligent molecular and atomic devices, leading to an unprecedented understanding and control over the basic building blocks of all natural and man-made things. Roco's vision encompasses the need for research that will have interaction between light and matter, a machine-human interface, and atomic manipulation to design molecules.

Molecularly assembled nanoweapons are based on molecular nanotechnology (MNT). MNT technology builds nanosized structures to complex, atomic specifications by means of mechanosynthesis. This is distinct from nanoscale materials. Based on Richard Feynman's vision of miniature factories using nanomachines to build complex products (including additional nanomachines), this advanced form of nanotechnology (or *molecular manufacturing*) would make use of positionally controlled mechanosynthesis, which is guided by molecular machine systems. MNT would involve combining physical principles demonstrated by biophysics, chemistry, and other nanotechnologies and the molecular machinery of life with the systems engineering principles found in modern macroscale factories. According to Eric Drexler, "On the molecular scale, you find it's reasonable to have a machine that does a million steps per second, a mechanical system that works at computer speeds."

Frietas has envisioned that varieties of nanomachines can be manufactured by MNT by using carbon-rich diamond-like materials, nanoparticles of aluminum-rich sapphire (Al_2O_3), boron-rich (BN) or titanium-rich materials; he has suggested that the carbon-based biomass can also serve as a power source for nanomachine replication [31].

Fabrication of nanoweapons by assembling molecules can be designed to be used as a Trojan horse to deliver or inject lethal toxin doses of hyper-destructive chemicals that are lethal at nanogram level. A single small package of such a device can carry billions of such weapons to destroy a population of billions of human beings and other living things, including crops, etc., in one go.

5.7.1.2 *Mini-Nukes and Mosquito-Like Robot Weapons*

Mini-nukes and mosquito-like robot weapons are being considered for future warfare by several countries. Such nanoweapons could be used for firing mini-nuclear bombs and insect-like lethal robots. These insect-like nanobots could be weapons of mass destruction as they can be used or programmed to perform many tasks like injecting toxins into people or contaminating the water supply; they can act like a nano drone and fly into a room to drop poison onto food for targeting a particular group or individual.

Since nanoweapon research is mostly carried out in an army or defense laboratory, most of the information on their developments are highly secret projects. However, there have been reports that an advanced drone the size of a fly with a set of tiny robotic legs has been fabricated, which can enter any building undetected to perform their function [32].

Another horrific form of nanoweapon that is in the offing is *micro-explosives* delivered with nanobots, which may be bioweapons with inhalable crippling chemicals or toxins. Eventually these autonomous bots may become self-replicating.

5.7.1.3 Invisible Nano-Needle Bullets

Invisible bullets or needle bullets are other devices which are now at the fantasy level but may soon become a reality. This future weapon of mass destruction can be fired from a great distance to disable victims. Similarly, molecularly manufactured nanodust can be used as infiltrators to affect brain functions, i.e., through neurons to control the mind of the victim, or even may function as "nano mind erasers" that will modify or erase a person's memory by targeting specific brain areas [33].

Nano-diameter needles, invisible to the human eye, are non-lethal weapons that can shoot clusters of bullets from far distances to "pin" people to a wall or freeze their physical movement without leaving any visible wounds or causing any permanent damage. It is hoped that the precision which comes with the nanotechnological approach will cause less loss of life.

5.7.1.4 Non-Nuclear Bomb

As televised on Russian TV, the news is out that using nanotechnology Russians have built the world's most powerful non-nuclear air-delivered bomb. Russians have named it the "Father of All Bombs" and claimed that it is four times more powerful than the American "Mother of All Bombs," which is also a very powerful non-nuclear weapon. This bomb does not release radiation, hence does not pose an environmental threat. Though the Russian bomb contains less explosive (7 tons, which is equivalent to 44 tons of regular explosives) than the American bomb (8 tons but equivalent to 11 tons of TNT), it has become more powerful because it uses a new, highly efficient type of explosive (the type of explosive is not stated) developed with the use of nanotechnology. The Russians can claim one more feather in their cap, because their bomb had a blast radius of 300 meters, which was two times bigger than the American bomb; and the temperature in the epicenter of the Russian bomb's explosion was also twice as high as that of the American bomb.

Innovative research efforts for the future advancement of weapons, such as high-performance non-nuclear bombs, are being done by processing explosive formulations using nano-aluminum powder. It is expected to give enhanced lethality in miniaturized, reduced volume of advanced aircraft munitions bays and for weaponization of unmanned air vehicles (UAVs). Nanoparticles have much higher surface area and are therefore more reactive for a given weight than conventional powders. Moreover, the reaction rate of nanopowders are size dependent; hence, can be controlled by adjusting the nanoparticle diameter for optimized air blast and for fragment formation and acceleration. A combination of nano-aluminum powders and conventional munitions formulations is being pursued at the industrial level.

5.7.1.5 Nanoweapons to Replace or Improvise Current Nuclear Weapons

Use of nuclear weapons is a rather complex procedure that needs not only miniaturization but also replacement or improvisation.

One might think that nanotechnology will resolve unsolvable conflicts, because with nanoparticles taking over the demand for natural resources such as gold, oil, diamonds, and other precious metals, a technological resolution may be amenable. With a nano-approach, the world economies may shift to getting easily available natural resources and with incorporation of nanotechnology-based sophisticated tech-fabricated elements, much of the money will drop out of the areas of the battlefront. Such a possibility has been successfully demonstrated.

5.7.1.6 New Nano Spies – Nano-Sized Fighter Jets

Scientists are working on the idea of making nano-sized fighter jets of only a few mm in size that can go anywhere, both indoors and outdoors, carrying nanoweapons; and remote-controlled nano air vehicles that can carry a payload of up to one fifth their weight, which will be able to collect military intelligence and help protect the lives and enhance the operational effectiveness of soldiers and first responders [34, 35].

5.7.1.7 CornerShot

CornerShot is a weapon accessory invented by Lt. Col. Amos Golan of the Israeli Defense Forces in cooperation with American investors [36].

As the name suggests, CornerShot is a weapon which can fire around corners. It was designed in the early 2000s for use by SWAT teams. Fitted with a high-resolution digital camera that is provided with color LCD monitor, it provides a video observation and sighting system and transmission capability, and can operate during the day or at night. Several variants of CornerShot have been designed so that it can be used in many different weapons, such as a pistol with a 40 mm grenade launcher; as a surveillance tool. The standard CornerShot can be mounted on the front of a semi-automatic pistol which is linked to a trigger in the rear part. It is 820 mm long and weighs 3.86 kilogram. Nanotechnology using high strength, nanocrystalline metals and lightweight nanocomposites is expected to create a version of it with reduced weight. There are efforts to develop nanosensor systems using IR, acoustic sniper array, etc.

5.7.1.8 Laser-Guided Weapons

There are laser-guided target assessment weapons, e.g., South Korea's XK-11 rifle, that can guide explosive projectiles and have mini missile launchers. They can fire NATO standard 5.56 mm rounds and 20 mm grenades. Using nanotechnology is expected to help in developing directional, sweeping radar array systems. Moreover, due to the very high optical efficiency of quantum dots, quantum structures are expected to play an important role in generating and sensing microwave radiation more efficiently. Thus, smaller and lighter detection devices can be developed.

The Israeli defense contractor Rafael Advanced Defense Systems Ltd, known as Rafael, has developed a high-energy laser weapon system (using powerful, air cooled laser, measuring up to 700 watt) known as Thor. It can defeat improvised explosive charges (IEDs), roadside bombs and unexploded ordnance (UXO) and other explosives from a safe standoff distance. Thor has dual capability and can be used for both offensive and defensive purposes. It can also be used for safe standoff removal of explosive obstacles by laser-directed energy or projectile kinetic energy. Since the laser is used to burn the IED, rather than activate it, it can avoid collateral damage which can result from other neutralization procedures.

Another very advanced laser-operated weapon is the 30–100 kW (LLNL) solid-state heat-capacity laser developed by Lawrence Livermore National Laboratory.

This compact laser system can destroy a landmine or IED beam and excavate soil by creating micro-explosions in the ground water, burning other types of covering, such as canvas or vegetation; then the beam focuses on the device (it can heat a metal container or drill through a plastic container); finally, the high explosive is activated. Here also, nanotechnology using optical quantum structures can play a role in future laser systems that will be more efficient, smaller and lighter laser devices set up for a movable 100 kW solid-state laser system (LLNL) heat effect.

As announced by MSN News, India developed its first laser-guided bomb kit in 2006, which has an effective firing range of 9,000 m (9,800 yd). It is a maneuverable, free-fall weapon requiring no electronic interconnection to the aircraft. It is named *Sudarshan* and is manufactured by Bharat Electronics Limited (BEL).

5.7.1.9 Bullet Camera (TNO Concept)

The bullet camera is a cross between a bullet and a UAV (unmanned aerial vehicle). Miniaturized with the help of nanotechnology, the bullet is composed of two parts (i) a sensor system (payload) and (ii) the point of the bullet that consists of an impact absorbing material that offers crash protection for the payload during the impact. So, the bullet will work as an intelligent monitoring "node" that can detect and do the surveillance tasks. The bullet can be fired by a traditional weapon. Apart from the camera, the bullet can also carry a microphone, infrared or radar array, acoustic detector, chemical sensors, radio for rf-communication, a small battery and, of course, the explosive charge.

5.7.1.10 Landmines and Improvised Explosive Devices

Landmines and improvised explosive devices are needed for the disposal of unexploded ordnance. Such devices should have components to perform detection, exposure of the device (for example, by excavation), and neutralization of the device.

Nanotechnology components, such as quantum structures and array technology on a chip, make them smaller, more powerful and capable microwave generators that may be used for detection.

For exposing landmines, quantum structures will offer smaller and more powerful lasers that can be used from a remote distance.

Neutralizing the device nanomaterials will offer increased energetic power to create smaller but more powerful and better explosive power.

5.8 Defensive Measures to Combat Nanoweapons

Defensive technologies to combat nanoweapons are envisaged [37, 38] for the development of

(i) Active shield against nanoweapons by providing encapsulation, thus protecting adhesion to the surface;
(ii) Nano-layer hard shields to defend against abrasions or heat;
(iii) Strong shields to protect against impact;
(iv) Passive protection by avoiding contact, sieves with molecule size pores protecting penetration;
(v) Active defense system like nano-missiles or projectiles to fire against approaching mini-micro-nano vehicles or devices;
(vi) Nano-immune system and/or anti-viral nanobots that could be released in living beings to trigger the detection of attacking nanobots and seed the living system with protective immune-system nanobots [39];
(vii) Nano-robots within the body to act as preventive inoculation of dangerous microbes and act as a guard;

(viii) Strategies for faster information processing;
(ix) Using artificial system in war, rather than human beings;
(x) Preventive attack mechanism; and
(xi) Above all, an anti-nanoweapon and anti-ecophagic policy.

5.9 Risks Posed by Nanoweapons

Nanoweapons may be a greater threat than nuclear weapons because nanotechnology can create havoc not only in areas like battle technology for nanoweapons but also physical fitness as well as command and control through nano-biotechnology and nano-computing [40]. The great nanotechnologist Eric Drexler compared the potential threat of nanoweapons and described it as follows: "A [nuclear] bomb can only blast things, but nanomachines...could be used to infiltrate, seize, change, and govern a territory or a world."

Bostrom [37] has forecasted existential risks (i.e., risks that are both global in scope and terminal in intensity) due to deliberate misuse of nanotechnology, such as self-replicating "nanobots" that will uncontrollably consume organic matter, spread and destroy the earth. Consequently, life on Earth will be poisoned and burned, and sunlight may even be blocked from reaching the Earth. Frietas [31] has scrutinized the possible damage caused by the MNT devices that may replicate at exponential rate and disperse all over with very high velocity liberating energy and thermal pollution that will cause ecophagic impact on the entire surface of the biosphere, causing damage to the surface and terrestrial ecology. Nanotechnology will produce weapons of mass destruction (WMDs) for future warfare. The advantages of nanoweapons of mass destruction are that they will be more varied, more deadly, more available, cheaper to obtain and easier to hide [41].

5.10 Need for Preventive Anti-Nanoweapon and Anti-Ecophagic Policies

Globally, a large amount of money is being spent in the area of nanoweapon development. Many articles and reports are available on the possible damages that can be caused by the use of nanoweapons. Hence, it is imperative that serious attention be paid to anti-nanoweapon and anti-ecophagic policies that address these concerns. The problem with nanoweapons is that unlike the massive nuclear weapons manufactured in large plants, facilities that manufacture smaller nanoweapons will be difficult to track. Moreover, unlike nuclear weapons, nanoweapons do not require rare isotopes like uranium, plutonium, etc.

Frietas [31] has proposed three specific policies to save the global ecosystem by ecophagic nanotechnologies: (i) An immediate moratorium on all artificial life experiments; (ii) Continuous comprehensive infrared surveillance of Earth's surface by geostationary satellites to monitor the current biomass inventory and to detect any rapidly developing artificial hotspots; (iii) R&D efforts to counteract ecophagic replicators, including scenario building and threat analysis with numerical simulations, measure/countermeasure analysis, theory and design of global monitoring systems capable of fast detection and response, IFF (Identification Friend or Foe) discrimination protocols, and eventually the design of relevant systemic defensive capabilities and infrastructure.

There will also be a need to limit nanoweaponry testing.

5.11 Summary

In this chapter, nano-sized materials and their composites incorporated into high energy materials (HEMs), i.e., propellants, explosives and pyrotechnics, are presented. It has touched upon the various methods of their synthesis and characterizations. The role of some of the experiments done using nanoparticles of AP, Al, RDX, HMX, HNF, CL-20, Metal oxides and their use in HEMs are discussed. Application of such fabricated products for use in different nanoweapons and nanoweapon-related support system, depending on their performance (burning rate, propagation velocity, ignition time, flame temperature, peak pressure in closed vessel, detonation pressure, etc.), is also presented. This chapter can be summarized by a quote from Lewis Carroll — *Imagination is the only weapon in the war against reality.*

References

1. Drexler, K.E., *Engines of Creation: The Coming Era of Nanotechnology*, Anchor Press/Doubleday, New York, NY, 1986.
2. Navrozov, L., "Future Wars Will Be Waged With Nano-Weapons", newsmax.com, September 5, 2008. http://www.newsmax.com/navrozov/drexler_nanotechnology/2008/09/05/128018.html.
3. Kapoor, I.P.S., Shrivastava, P., Singh, G., *Propell. Explos Pyrotech*, 34, 4, 351–356, 2009.
4. Gromov, A., Strokova, Y., Kabardin, A., Vorozhtsov, A., Taipei, U., *Propell. Explos Pyrotech*, 31, 6, 452–455, 2009.
5. Pivkins, A., Frolov, Y., Evanov, D., Meerov, D., Monogarov, K., Nikolaskya, A., Mudestova, S., Plasma synthesized nano-aluminum powder: Thermal properties and burning with ammonium perchlorate. *Proceedings of 33rd International Pyrotechnics Seminar*, Fort Collins, Colorado, July 16-21, pp. 141–151, 2006.
6. Huang, H., Wang, J., Xu, W., Xie, R., *Propell. Explos. Pyrotech.*, 34, 1, 78–83, 2009.
7. Jones, D.E.G., Turcotte, R., Fouchard, R.C., Kwok, Q.S.M., Vacon, M., *34th International Annual Conference of ICT*, Karlsruhe, Federal Republic of Germany, June 24 to 27, pp. 33-1 to 33-15, 2003.

8. Jones, D.E.G., Turcotte, R., Kwok, Q.S.M., Vacon, M., Turcotte, A.M., Abdal-Qadar, Z., *Propell. Explos. Pyrotech.*, 29, 30, 120–131, 2003.
9. Breussau, P. and Anderson, C.J., *Propell. Explos. Pyrotech.*, 27, 5, 300–305, 2002.
10. Tappan, B.C. and Brill, T.B., Cryogel synthesis of nano stabilized CL-20, coated with cured nitrocellulose. *34th International Annual Conference of ICTKarlsruhe, Federal Republic of Germany*, June 24-27, pp. 29-1 to 29-10, 2003.
11. Gagulya, M.F., Makhov, M.N., Dolgoborodov, A.Y., Brazhnikov, M.A., Leipunsky, I.O., Jigatch, A.N., Kuskov, M.L., Laritchev, M.N., Aluminium nano composites based on HMX. *34th International Annual Conference of ICTKarlsruhe, Federal Republic of Germany*, June 24 to June 27, pp. 66-1 to 65-11, 2003.
12. Kanel, G.I., Utkin, K.V., Razoneronv, S.V., *Cent. Eur. J. Energ. Mat.*, 6, 1, 15–30, 2009.
13. Stamatis, D., Jiang, X., Beloni, G., Dreizin, E.L., *Propell. Explos. Pyrotech.*, 35, 3, 260–267, 2010.
14. Kosanke, B.J., Jennings-White, C., Kosanke, K.L., Pyrotechnic reference series No. 4. *Journal of Pyrotechnics* Inc, Whitewater, CO, USA.
15. Barbee, T.W., Gasj, A.E., Satcher, J.H., Jr., Simpson, R.L., Nanotechnology based environmentally robust primer. *34th International Annual Conference of ICT*, Karlsruhe, Federal Republic of Germany, June 24-27, pp. 31-1 to 31-13, 2003.
16. Appleyerol, F.G. and Nigel, D., Modelling infrared extinction of highly conducting micro and nano scale fibers. *Proceedings of 31st International Pyrotechnics Seminar*, Fort Collins, Colorado, July 11-16, pp. 273–274, 2004.
17. Shekhar, H., Nano sized ingredients of propellants, explosives and pyrotechnics, in: *Nanotechnology Vol. 5: Defence Applications*, N.K. Navani, S. Sinha, ED Exec, J.N. Govil (Eds.), Studium Press LLC, USA, 2013.
18. Luo, Q., Long, X., Pei, C., Nie, F., Ma, Y., *J. Ener. Mater.*, 20, 2, 150–161, 2011.
19. Yang, G., Nie, F., Huang, M., Zhao, L., Pang, W., *J. Ener. Mater.*, 29, 4, 281–291, 2006.

20. Yongxu, Z., Dabin, L., Chunxu, L., *Propell. Explos Pyrotech*, 30, 6, 438–431, 2005.
21. Bayat, Y., Eghdamtalab, M., Zenali, V., *J. Ener. Mater.*, 28, 4, 273–284, 2010.
22. Nafday, O.A., Vaughn, M.W., Weeks, B.L., *J. Chem. Phys.*, 125, 144703, 2006.
23. Luo, Q., Long, X., Pei, C., Nie, F., Ma, Y., *J. Energetic Mater.*, 29, 2, 150–161, 2011.
24. Shokrolahi, A., Zali, A., Mousaviazar, A., Keshavarz, M.H., Hajhashemi, H., *J. Ener. Mater.*, 29, 2, 115–126, 2011.
25. Mang, J.T. and Hjelm, R.P., Characterization of components of nano energetic by small angle scattering technique. *Proceedings of 31st International Pyrotechnics Seminar*, Fort Collins, Colorado, July 11-16, pp. 299–305, 2004.
26. Lee, M., Advanced penetrator materials. *Armaments for the Army Transformation Conference,* NDIA Conference & Exhibition, Parsippany Hilton, Parsippany NJ, June 18-20, 2001.
27. Perram, G., Marciniak, M., Goda, M., High energy laser weapons: Technology overview. *Proc. SPIE 5414, Laser Technologies for Defense and Security*, September 2004, doi: 10.1117/12.544529.
28. Bonadies, G., Nano-weapons: Tomorrow's Global Security Threat, November 2008, doi: 10.13140/RG.2.2.31676.23684.
29. Roco, M.C., *J. Nanopar. Res.*, 7, 6, 707–712, 2004.
30. Roco, M.C., *AIChE J.*, 50, 5, 896, 2011.
31. Frietas, R.A., Jr., Some Limits to Global Ecophagy by Bivorous Nanoreplicators, with Public Policy Recommendations. 2, 2000. (http://www.rfrietas.com/Nano/Ecophacy.htm).
32. Sheftick, G., Army Developing Robotic Insects? Army News Serv., December 2014.
33. Bennett-Woods, D., *Nanotechnology: Ethics and Society*, p. 158, CRC Press, Boca Raton, FL, 2008.
34. Nanotechnology in the Military. http://uk.gizmodo.com/2006/07/21/nano_spy_plane_gets_green_ligh.html.
35. http://hsdailywire.com/single.php?id=6234.
36. Gibson, K., Bad news for the bad guys: The next best thing to being there! *Guns Mag.*, 2004.
37. Bostrom, N., *J. Evol. Technol.*, 9, 1, 1–30, 2002.

38. Altmann, J., *Nanotechnology and preventive arms control*, p. 34, University of Dortmund, Germany, 2005, http://www.bundessf tungfriedensforschung,de/pdf-docsberichtaltmann.pdf.
39. Frietas, R.A., Jr., *J. Comput. Theor. Nanosci.*, 2, 1–25, 2005.
40. Treder M. Nano-Guns, Nano-Germs, and Nano-Steel. A Review of Ultraprecision Engineering and Nanotechnology. *Nanotechnology Perceptions*: 2(1): 25-26, 2006.
41. Del Monte, L.A., *Nanoweapons: A Growing Threat to Humanity*, Potomac Books; An Imprint of the University of Nebraska Press, Lincoln, USA, 2017.

6

Nanotechnology to Aid Biological and Chemical Warfare Defense

Madhuri Sharon

Walchand Centre for Research in Nanotechnology & Bionanotechnology, WCAS, W.H. Marg, Ashok-Chowk, Solapur, Maharashtra, India

The question is, what levels of insanity do we have to prepare for?
Joshua Lederberg
Nobel laureate

6.1 Introduction

The destructive nature of mankind evolved from the need to survive. Right from the beginning of civilization, primitive man used rocks, wood and later metals to hunt for food and to kill for survival. Later, this power of weapons transformed into lust for more and more; and this tendency through a series of developments of various weapons, has eventually resulted in today's biological and chemical warfare.

There has been immense development in catapults and siege machines, which are now also being used for biological and chemical warfare. Present-day chemical and biological weapons have seen various phases of development, for example,

Madhuri Sharon, et al. Nanotechnology in the Defense Industry: Advances, Innovation, and Practical Applications, (165–234) © 2019 Scrivener Publishing LLC

(i) During World War I, gaseous chemicals like chlorine and phosgene were used in the Battle of Ypres; (ii) During World War II, chemical nerve gas (tabun, a cholinesterase inhibitor) and biological warfare, such as the use of *Bacillus anthracis* for inflicting anthrax, *Salmonella* for contaminating water, *Vibrio cholerae*, *Shigella* spp. and *Yersinia pestis* for causing epidemic diseases, and the release of fleas infected with plague (used by the Japanese against the Chinese) came into existence; (iii) In 1970, the Vietnam War ushered in another phase with the use of lethal chemicals like Agent Orange and a mix of crop-damaging and tree-defoliating herbicides; (iv) In the recent era, knowledge of biotechnology and genetic engineering is being used for advancing the development of bioweapons like genetically designed hazardous organisms that can produce harmful chemicals (toxins, venoms, bioregulators), which are resistant to antibiotics, routine vaccines and therapeutics. Such organisms alter unknown immunologic profiles so they cannot be diagnosed and may escape detection by antibody-based sensors.

6.2 What Is Biological Warfare?

In a nutshell, biological warfare is the use of hazardous material of biogenic origin, such as toxins, and even infectious microbes like bacteria, virus, fungi or harmful insects. These microbes can kill all living beings, including plants. Biological materials that are used as weapon are called bioweapons, biological agents or sometimes weapons of mass destruction (WMDs). These biological weapons can kill an entire population or destroy vegetation. There have been umpteen cases of biological war in the past (Table 6.1).

However, the use of biological weapons is now considered a war crime [12] and is prohibited under International Humanitarian Law. According to the 1972 Biological Weapons Convention, it is suggested that the international community should adopt

Table 6.1 Biological war practiced in the past.

When	By whom	Biological agents used	References
1500–1200 BC	Assyrians	Poisoned enemy wells with ergot (a poisonous fungus), causing a deadly epidemic known as the Hittite Plague.	Trevisanato [1]
600 BC	Athenian legislator Solon	Contaminated the Pleisthenes River with the roots of Helleborus plants to give his enemies severe diarrhea, leading to their defeat.	Mayor [2]
300 BC	Greeks	Polluted wells used by their enemies for drinking water with animal corpses.	Southern Illinois Univ. Report 2012
200 BC	Carthaginian General Maharbal	Left behind a large stock of wine that he treated with the toxic roots of the Mandragora plant which causes a narcotic effect. The enemy, upon drinking the wine, fell into deep sleep and the Carthaginians returned to slay them.	Rothschild [3]

(Continued)

Table 6.1 Biological war practiced in the past. (*Continued*)

When	By whom	Biological agents used	References
184 BC	Carthaginian soldiers led by Hannibal	Catapulted pottery jars filled with poisonous snakes onto the decks of enemy ships in the battle against King Eumenes II of Pergamum to achieve victory.	Rothschild [3]
6th Century BC	Assyrians	Poisoned wells with fungus that rendered the enemy delirious.	US Biological warfare Program History
1346	Mongol Tatars	Bodies of dead Mongol warriors (who died due to Plague) were used to infect and kill people of the Crimean City of Kaffa (now Feodosia, Ukraine)	Wheelis [4]
1763	British Army	Gave blankets carrying the smallpox virus to Native Americans in Ohio during the Siege of Fort Pitt in 1763.	Calloway [5]

(*Continued*)

Table 6.1 Biological war practiced in the past. (*Continued*)

When	By whom	Biological agents used	References
1789	British Marines	Deployed the smallpox virus as a means of defense against indigenous tribes in New South Wales.	Warren [6]
1914–1918 World War I	Imperial German Government	Used bioagents in the form of anthrax and glanders.	Baxter & Buergenthal [7]
1939–1945 World War II	Imperial Japanese Army Unit 731	Biological weapons were used against both Chinese soldiers and civilians.	Hal Gold [8]
1940	Japanese Army Air Force	Bombed Ningbo with ceramic bombs full of fleas carrying the bubonic plague.	Barenblatt [9]
1952 Mau Mau Uprising in Kenya	Mau Mau	Killed cattle with the poisonous latex of the African milk bush plant.	Lockwood [10]
1962–1971	U.S. Air Force	Sprayed 11 million gallons of the herbicide 2,4-D to defoliate the forest trees where Vietnamese fighters used to hide.	U.S. Military and Herbicide Program in Vietnam [11]

the Convention on the Prohibition of the Development, Production and Stockpiling of Bacteriological (Biological) and Toxin Weapons and their Destruction.

However, clandestine efforts to conduct biological warfare do exist. It involves keeping the biological agent alive and virulent, transporting it to the place of delivery and dispersing it. For example, anthrax spores can be sprayed by an aircraft using a pesticide sprayer, or delivered by mail, as was previously done. Culture tubes containing microbes, like *Salmonella*, can be transported and added to food. Biological weapons can be employed against humans, animals or even plants. These weapons can be used for individual assassination, for small-scale targeted attacks, or even for mass destruction.

6.2.1 Types of Biological Warfare

The requirements for being an effective biological agent include high morbidity and mortality, easy to multiply, transmit and spread, infective at low dose, and highly infectious by aerosol. Selection of biological agent is done based on its environmental stability, lack of rapid diagnostic system, lack of vaccine, feasibility of production and, above all, potential to be weaponized. Potential to be weaponized means the possibility of either using microbes in natural form or genetically modifying them for maximum desired impact such as antimicrobial resistance, evasion by the immune system, creation of fine-particle aerosols or alteration of host range through changes in surface proteins and enhancing the toxin production to cause more devastating effect.

Some of the biological materials used in biological warfare are briefly discussed below.

6.2.1.1 Bacteria

Types of bacteria used for weaponization are mentioned below.

6.2.1.1.1 *Bacillus anthracis*

Bacillus anthracis (Figure 6.1) is the Gram positive, rod-shaped microbe (1–1.2 μm in width and 3–5 μm length) that causes anthrax disease in both humans and animals. Its endospores (Figure 6.2) are perfect for dispersal aerosols. The endospores are formed only under aerobic conditions. Spores can survive in soil for decades. They can enter through the skin, GI tract and lungs, where they incubate within 3–4 days; initially they cause flu-like symptoms and within three to seven days develop into

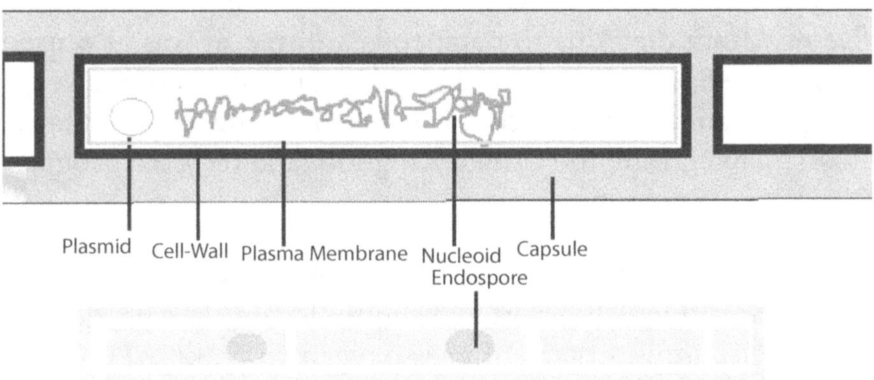

Figure 6.1 Schematic diagram of *Bacillus anthracis.*

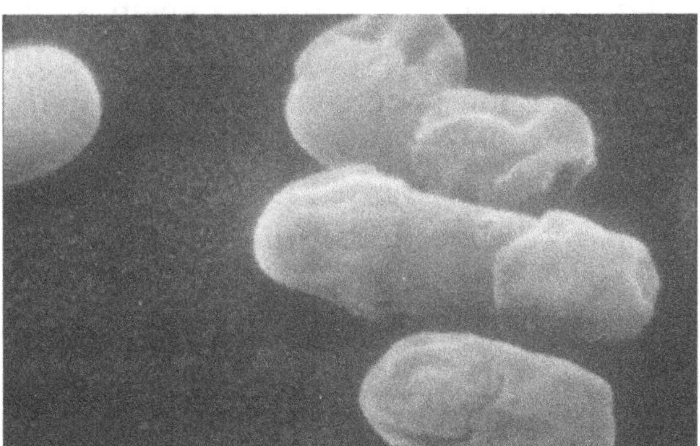

Figure 6.2 SEM image of *Bacillus anthracis* spores.

a fatal hemorrhagic mediastinitis (inflammation of the tissues in the mid-chest). The fatality rate with this infection is > 90%. It is a relief that it is not a transmissible infection and can be treated with intravenously given β-lactam antibiotic (penicillin, doxycycline, tetracycline and ciprofloxacin). Moreover, there is an anthrax vaccine, Anthrax Vaccine Adsorbed (AVA), that protects against cutaneous and inhalation anthrax. The requirement is early detection of the disease.

There is a record of bioterrorism using spores of Ames strain of *Bacillus anthracis,* sent by post on January 26, 2002, to six people in the United States. Eleven people inhaled the anthrax, five of whom died due to cutaneous anthrax. It was also used during the Second World War. Anthrax is a focus of bioweapons programs by many countries such as the United States, Russia, Japan, Iran, the United Kingdom and terrorist groups.

6.2.1.1.2 *Brucella suis*

Brucella suis is a Gram-negative coccobacillus. It is a facultative organism which grows and multiplies inside a phagocytic cell. A *facultative* organism uses oxygen to make ATP; when it is not available it "*exercises its option*"—the literal meaning of the term—and makes ATP via fermentation or by substituting one or more of four less efficient electron acceptors as oxygen at the end of the electron transport chain: sulfate, nitrate, sulfur or fumarate. *Brucella suis* infects pigs and through the food supply chain it infects humans and causes undulant fever, which in infected pregnant females causes an abortion of the fetus. Once infected, the relapse rate is very high. Apart from *Brucella suis; B. melitensis, B. abortus, B. ovis,* and *B. canis* are also recognized as potential agricultural, civilian, and military bioterrorism agents [13]. *B. melitensis* infects goat and sheep; *B. abortus* infects cattle and *B. suis* infects pigs, but all three of these species can infect humans. For therapy, a combination of streptomycin and tetracycline or rifampicin is given. The United States tested *B. suis* as the first biological weapon agent in 1952. No Brucella vaccine has been developed for the protection of humans. However, there are several licensed live Brucella vaccines for livestock.

6.2.1.1.3 *Burkholderia mallei* and *Burkholderia pseudomallei*
Burkholderia mallei and *Burkholderia pseudomallei* are closely related species. They are Gram-negative, bipolar, aerobic and non-motile bacteria found in soil and water. Their size varies from 1.5–3.0 μm in length and 0.5–1.0 μm in diameter. It is an obligate mammalian microbe and is transmitted from one host to another. Horses are the primary hosts of these bacteria, but it can also affect mules, donkeys, goats, dogs, cats and even humans. It causes glanders and melioidosis, which are both highly infectious diseases. Symptoms of glanders are fever with chills and sweating, headache, muscle aches and tightness, chest paint, running nose and light sensitivity. It is considered for biological warfare because only a few germs of *B. mallei* are enough to trigger the disease. Melioidosis starts with chest infection, its symptoms occur 2–4 weeks after exposure and include fever, headache, loss of appetite (anorexia), cough, shortness of breath, pleuritic chest pain, and general muscle soreness. This disease was noted in many soldiers during the Vietnam War. To diagnose melioidosis, complete detailed screenings are required, i.e., blood culture, sputum culture, urine culture, throat swab, and culture of any aspirated pus.

It can be disinfected by many disinfectants, such as benzalkonium chloride, iodine, mercuric chloride, potassium permanganate, 1% sodium hypochlorite, ethanol, etc., and can be destroyed thermally or by UV irradiation. Since melioidosis has a possibility of recurrence, the treatment of melioidosis is divided into two stages: an intravenous high-intensity phase and an eradication phase to prevent recurrence. Amikacin, amoxicillin/clavulanic acid, carbapenems, ceftazidime, chloramphenicol, doxycycline, piperacillin, streptomycin, tetracycline and sulfathiazole have been found to be effective against this bacterium when tested *in vitro*. The eradication treatment is done with co-trimoxazole and doxycycline for 12 to 20 weeks to reduce the rate of recurrence [14]. So far, no vaccine has been developed against this microbe. *B. mallei* and *B. pseudomallei* have been classified by the Centers for Disease Control and Prevention as category B critical biological agents because they are highly infective. They have been used in World War I. They are in the biological warfare program of the United States, Russia and Japan. There is a report that they have been used during the war in Afghanistan against the Mujahideen [15].

6.2.1.1.4 *Chlamydophila psittaci*

Chlamydophila psittaci is a Gram-negative obligate intracellular, coccoid and non-motile bacterium (Figure 6.3). Its size ranges from 0.2–1.5 m and the cell wall lacks peptidoglycan, instead it has an outer membrane containing lipopolysaccharide and a cytoplasmic membrane bilayer [16]. In humans it causes a disease known as psittacosis or parrot fever.

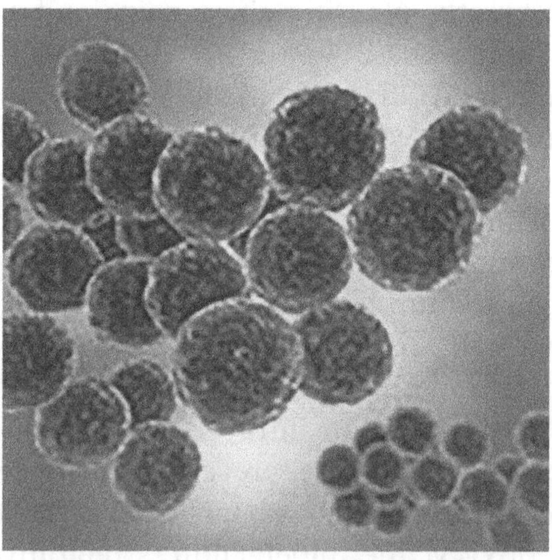

Figure 6.3 *Chlamydophila psittaci.*

Its main hosts are feral birds and poultry; however, it also infects cattle, pigs, sheep and horses. In mammals it causes endemic avian-chlamydiosis, epizootic outbreaks; in humans and birds it causes respiratory psittacosis, which develops into fatal pneumonia. *Chlamydia psittaci* is transmitted by inhalation, contact or ingestion. Birds are the vector of chlamydia infection. There are eight different serovars of *C. psittaci*. When inhaled, the spherical elementary body of *C. psittaci* can penetrate human lung cell by phagocytosis or receptor-mediated endocytosis. Once inside the cell, it enlarges and becomes reticulate, where it replicates into many spherical elementary bodies and then is released by rupturing the cell and entering into other host cells for further multiplication [17, 18]. Thus, by using energy from infected cells it multiplies very fast. The source of *C. psittaci* in humans is usually the discharge coming from beaks, eyes, feces, and urine that contaminates the feathers of an infected bird. An infected person shows pneumonia- and or typhoid-like symptoms. Intravenous injection or oral administration of tetracycline hydrochloride or doxycycline [19] is used for treating psittacosis in adults, whereas children less than eight years old are treated with erythromycin. No human vaccine is available for psittacosis. The mortality rate for untreated psittacosis is from 15–20%.

C. psittaci was considered a possible biological warfare threat after an outbreak of psittacosis became an epidemic from 1920–1929, when a shipment of Amazon parrots from Argentina was shipped to many parts of the world. Due to this epidemic, many countries banned the import of birds. Psittacosis is labeled as a category B biological weapon agent that can be a possible incapacitating agent.

6.2.1.1.5 *Coxiella burnetii*

Coxiella burnetii is a Gram-negative, intracellular obligate pathogenic bacterium (Figure 6.4). *C. burnetii* can survive in high temperature, osmotic pressure, under ultraviolet light

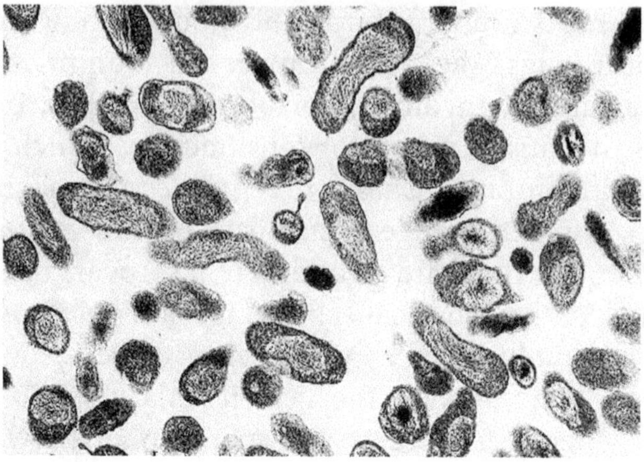

Figure 6.4 *Coxiella burnetii.*

and standard disinfectants. Infected cattle, goats, and sheep serve as a reservoir for the bacteria. If inhaled it causes Q fever in humans. Its incubation period is 9–40 days. Symptoms of Q fever are like the flu, with abrupt onset of fever, malaise, profuse perspiration, severe headache, muscle pain, joint pain, loss of appetite, upper respiratory problems, dry cough, pleuritic pain, chills, confusion, and gastrointestinal problems such as nausea, vomiting and diarrhea. It can progress to cause acute respiratory distress syndrome, which can be fatal. Suitable antibiotics for Q fever therapy are tetracycline, doxycycline, chloramphenicol, ciprofloxacin, ofloxacin, hydroxychloroquine or a combination of erythromycin and rifampin. Q-VAX vaccine (CSL) is an effective preventive measure against Q fever.

C. burnetii was developed as a potential biological weapon in the 1950s [20] and tested on volunteers as well as under desert conditions. *C. burnetii* was one of seven agents that have been standardized as biological weapons [21]. It is currently ranked as a category B bioterrorism agent by the CDC [22] because it is very contagious, is very stable in aerosols in a wide range of temperatures, can survive on surfaces up to 60 days, and its

ID_{50} (number of bacilli needed to infect 50% of individuals) is considered to be one, making it the lowest known.

6.2.1.1.6 *Escherichia coli* 0157:H7

Escherichia coli, also known as *E. coli,* is a Gram-negative, facultative anaerobic, rod-shaped, coliform flagellate having peritrichous arrangement, motility and nonsporulating bacterium of about 2.0 µm long and 0.25–1.0 μm in diameter (Figure 6.5). *Escherichia coli* is a common flora of the gut that helps in producing vitamin K. However, the presence of pathogenic serotypes of *E. coli* in the lower intestine of warm-blooded organisms causes serious food poisoning, where it grows in the fresh fecal matter under aerobic conditions for 3 days, but its numbers decline slowly afterwards. Cells that come out with feces can survive outside for a limited amount of time. Transmission is via the fecal–oral route.

Escherichia coli O157:H7 is a serotype of the species *E. coli* that produces Shiga toxin, which causes "colonic escherichiosis" through consumption of contaminated and raw food, including raw milk. This leads to hemorrhagic diarrhea, abdominal cramps, fever and in some cases kidney failure. In people with weak immune system, the infection can cause hemolytic uremic syndrome (HUS), in which the red blood cells are destroyed and the kidneys fail.

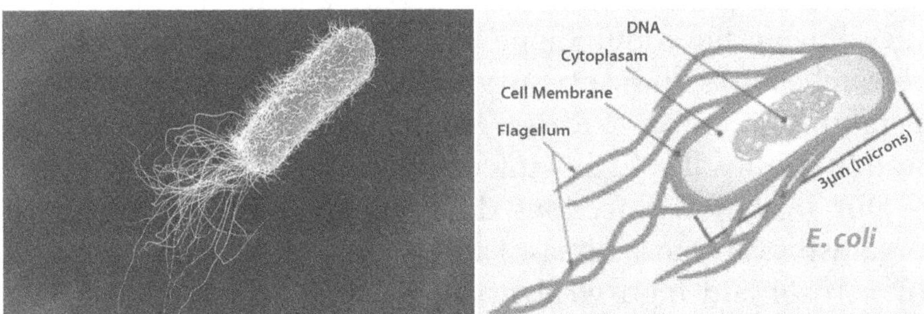

Figure 6.5 SEM image and schematic of *E. coli.*

A stool culture can diagnose the infection. To prevent the death fluid replacement, blood pressure support and anti-Shiga toxin antibodies are recommended.

The malleability of *E. coli* is a potentially deadly biological weapons agent. In one trial by American scientists in the 1980s, a gene that produced lethal protein in anthrax was inserted into harmless *E. coli*, which successfully produced the same deadly proteins as anthrax. The Iraqi biological weapons program has also demonstrated interest in the possibilities of altered *E. coli* bacteria as a biological agent, and shipments of this *E. coli* to Iraq from the United States have been reported.

6.2.1.1.7 *Francisella tularensis*

Francisella tularensis is a Gram-negative rod-shaped coccobacillus, an aerobic bacterium of about 0.2 by 0.7 μm in size. It is a non-spore-forming, non-motile, fastidious, facultative intracellular pathogenic bacterium. *F. tularensis* has been found in birds, reptiles, fish, invertebrates, and mammals, including humans. It can enter through blood or the respiratory system, due to consumption of contaminated food, by fly bites and conjunctival infection of the eye. Biovar of *F. tularensis* causes lethal pneumonia. It is taken up by phagocytosis into phagosomes where it multiplies; then the phagosomes burst and release them to cytosol so as to infect other cells and cause apoptosis. But human to human transmission has not been reported. The antibiotics used to treat tularaemia are amino-glycosides, tetracycline, or fluoroquinolone. A vaccine development is being researched. It is known to survive for several weeks at low temperatures in animal carcasses, soil, and water. *F. tularensis* can lead to fatality. It is highly infectious bacteria.

Due to its low infectious dose (inhalation of as few as 10 organisms can cause disease) and the ease at which it is spread by aerosol and high virulence, *F. tularensis* is classified as a Tier 1 Select bioterrorism agent by the U.S. government. In the past, many countries developed this bacterium into a biological weapon for aerosol dissemination. In 1970, a World Health Organization (WHO) expert committee reported that if 50 kg (110 pounds) of virulent *F. tularensis* was dispersed as an aerosol over a metropolitan area with a population of 5 million, there would be an estimated 250,000 incapacitating casualties, including 19,000 deaths.

6.2.1.1.8 *Rickettsia prowazekii* and *Rickettsia rickettsii*

Rickettsia prowazekii is a small, Gram-negative, obligate intracellular, rod-shaped bacterium (Figure 6.6). *R. prowazekii* grows freely in the cytoplasm of infected host cells and infects endothelial cells in humans. *R. prowazekii* causes epidemic typhus, which is transmitted by the defecation of body louse (*Pediculus humanus corporis*) that feeds on the blood of the patients infected with primary epidemic typhus. When the host scratches the wound, it contaminates the wound with *R. prowazekii*-infected feces. Close personal contact or sharing of clothes spreads the lice from person to person. Doxycycline or chloramphenicol is the therapeutic agent for *R. prowazekii.*

R. rickettsii rapidly die after killing their host cells, whereas *R. prowazekii* can retain high aerosol infectivity and viability in dried louse feces and remains highly infectious following drying in media with high osmomolarity; therefore, it has been weaponized as an agent of bioterrorism. Among the 17 named

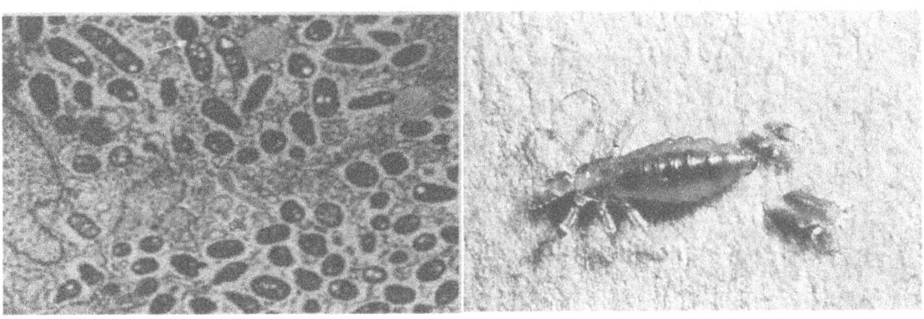

Figure 6.6 (left) TEM of *Rickettsia prowazekii* and (right) adult female human body louse and 2 larvae.

species of the Rickettsiaceae that have been documented to cause human illness, 5 are significant threats of a fatal outcome, namely *Rickettsia rickettsii*, *Rickettsia prowazekii*, *Orientia tsutsugamushi*, *Rickettsia typhi*, and *Rickettsia conorii*. These have infectivity by a low dose. Thus, they could be used by terrorists. Epidemic louse-borne typhus requires a vaccine to control large epidemics.

6.2.1.1.9 *Salmonella* spp.

Salmonella species are rod-shaped, Gram-negative, non-spore-forming, motile enterobacteria having 0.7 to 1.5 μm diameter and 2 to 5 μm length, and peritrichous flagella (all around the cell body). They are chemotrophic, facultative anaerobic, intracellular pathogenic bacteria. Out of 2,500 strains of bacteria, more than 1,400 serotypes are pathogenic strain of *Salmonella bongori*. According to the World Health Organization (WHO), there are three categories of *Salmonella*, i.e., typhoidal, non-typhoidal and animal. Low temperature of 0–4 °C arrests the growth of *Salmonella* bacteria but does not kill them, whereas high temperature of 60 °C deactivate the bacteria. They usually infect the gastrointestinal tract and cause food poisoning. If they invade the blood stream, some strains are invasive and cause typhoid fever, which requires antibiotic treatment. After invading the blood stream, they spread throughout the body and secrete a septic form of endotoxins, causing fatal hypovolemic shock and septic shock.

After the report given by the U.S. Centers for Disease Control and Prevention (CDC) in 2006 that *Salmonella*-related infections cause an estimated 1.4 million cases of foodborne illness and approximately 500 deaths annually in the United States alone; it has been thought to be a potential biological agent, as it can be easily weaponized as aerosol or contaminated food and water.

The biological weapon program of France during World War I and World War II, Japan's infamous Unit 731 activity and

South Africa's Project Coast have studied the potential use of *Salmonella* bacteria as a biological weapons agent. Terrorist groups have also found use of *Salmonella* bacteria as an easily accessible biological agent. In September 1984, in an attempt to influence local elections in Antelope, Oregon, it was suspected that the Rajneesh cult contaminated local salad bars with *Salmonella enterica* in an attempt to suppress the vote.

6.2.1.1.10 *Shigella* spp.

Shigella dysenteriae is a well-known food borne pathogenic bacteria. *Shigella* is a non-spore forming, non-motile, rod-shaped Gram-negative, facultative, anaerobic bacterium (Figure 6.7), which aids in the facilitation of intracellular pathogens. It causes shigellosis, an infectious disease that affects the intestinal tract.

When infected food is consumed, *S. dysenteriae* passes through the acidic medium of the stomach and infects the intestinal tract. As few as 10 cells of *S. dysenteriae* are enough to cause infection in humans; its serogroup A can cause deadly gastroenteritis epidemics. Another species, *S. boydii* serogroup C (found in the Indian subcontinent), *S. flexneri* serogroup B (found in developing countries) and *S. sonnei* serogroup D (found in developing) and *S. flexneri* (distributed worldwide)

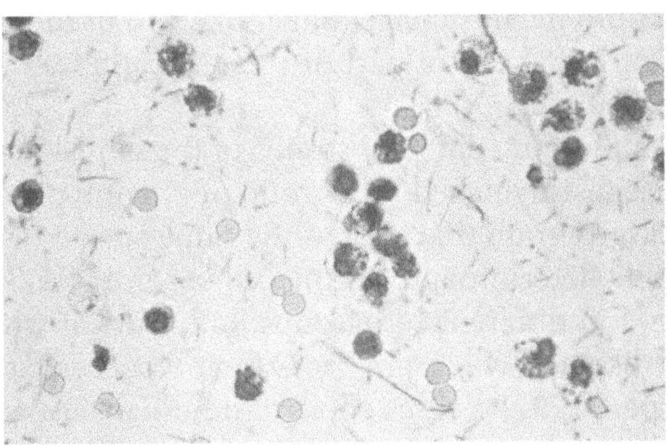

Figure 6.7 *Shigella dysenteriae.*

are endemic to specific regions. It can be transmitted through person-to-person contact, eating food contaminated by flies, drinking contaminated water and from swimming in contaminated water. Infected humans experience diarrhea (often bloody), fever, stomach cramps, pains in their joints, irritation of the eyes, painful urination and severe dehydration. Its incubation period is 1–4 days. Most people will recover from mild cases of shigellosis without antibiotic treatment. However, the antibiotics used for treatment of severe infections are ampicillin, trimethoprim/sulfamethoxazole, nalidixic acid and ciprofloxacin. *Shigella* infection is diagnosed by testing the stool.

Shigella has been used as a biological weapon because it is antibiotic resistant, highly infectious and there is no vaccine against it. It was used by the Japanese in Manchuria during World War II. Iraq has also developed it as a possible offensive biological agent and terrorists are also using it. As a bioweapon it can be spread by aerosols, contaminated water or by introducing contaminated food.

6.2.1.1.11 *Vibrio cholerae*

Vibrio cholerae is a Gram-negative, comma-shaped (curved rod), facultative, anaerobic, 0.5–0.8 μm wide and 1.4–2.6 μm long bacterium having a flagellum at one end of the cell as well as pili (Figure 6.8). It is a part of the natural flora of brackish water and estuaries.

It has both biovar (different strains of the same bacterial species that are distinguished by a group of phenotypic or genetic traits) and serogroup (bacteria of the same species with different antigenic determinants on the cell surface). *V. cholera* has more than 150 different serogroups, only two of which cause epidemic disease, which are (i) O1 biotype El Tor N16961, (ii) serotype O139. It causes cholera in humans. Apart from *V. cholerae* there are many other species of *Vibrio* associated with human health (Table 6.1).

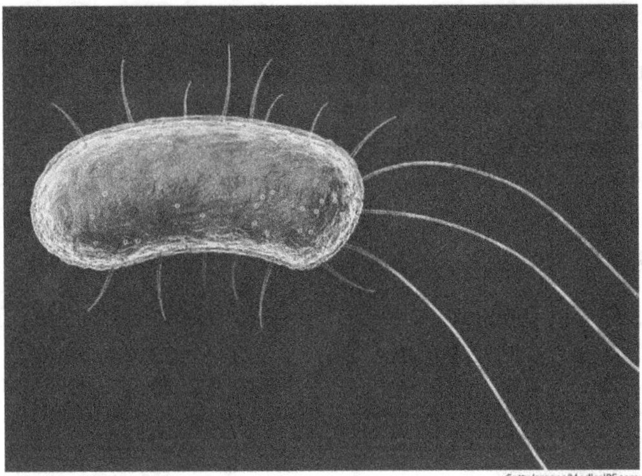

Figure 6.8 SEM image of *Vibrio cholera.*

Cholera is an infectious disease of the human intestinal tract that is caused by the ingestion of food or water contaminated by *Vibrio cholerae*. Its incubation period is 1–3 days. The patient suffers from profuse watery diarrhea and rapid dehydration, leading to a state of fatal collapse within hours if not treated. For treatment (oral or intravenous) rehydration of fluid and electrolytes is done [23, 24]. Effective drugs for treatment are doxycycline or tetracycline and vaccine.

V. cholerae can survive for 2–6 weeks in water contaminated with organic matter or sewage waters and up to 16 days in soil, and can resist freezing temperatures. However, it cannot survive heat, sunlight, and dry conditions [25, 26] and chlorine [27].

V. cholerae was used by the Japanese Army during World War II to contaminate the food and water supplies of Chinese cities [28] and is thought to have been utilized as a terrorist weapon to contaminate potable water [29].

6.2.1.1.12 *Yersinia pestis*

Yersinia pestis is a Gram-negative, facultative, anaerobic, non-motile, rod-shaped coccobacillus bacterium (Figure 6.9)

Table 6.1 Various species of *Vibrio* and the disease they cause in human.

***Vibrio* species**	**Source of infection**	**Disease**
V. cholerae	Water, Food	Gatroenteritis
V. parahaemolyticus	Shellfish, Seawater	Gastroenteritis, Bacteremia, Wound infection
V. vulnificus	Shellfish, Seawater	Bacteremia, Wound infection, Cellulitis
V. alginolyticus	Seawater	Wound infection, External Otitis
V. bollisae	Shellfish	Gastroenteritis, Wound infection, Bacteremia
V. fluvialis	Seafood	Gastroenteritis, Wound infection, Bacteremia
V. damsela	Seawater	Wound infection
V. metschnikovii	Unknown	Bacteremia
V. mimicus	Freshwater	Gastroenteritis. Wound infection, Bacteremia
V. furnissii	Seawater	Gastroenteritis
V. cincinnatiensis	Unknown	Bacteremia, Meningitis
V. carchariae	Seawater	Wound infection (by shark bite)

which does not form spores. It infects humans via the fleas of small mammals, such as rats, and causes plague, a usually fatal disease [30]. Plague is divided into three main types: pneumonic, septicemic and bubonic [31].

There are various medium where *Y. pestis* can survive, e.g., in water for 16 days, in moist soil for > 60 days [25]; and it can also survive for some time in dry sputum, flea feces, and buried bodies [26]. It can be killed by exposure to sunlight and high temperature. For chemical inactivation of *Y. pestis,* 1%

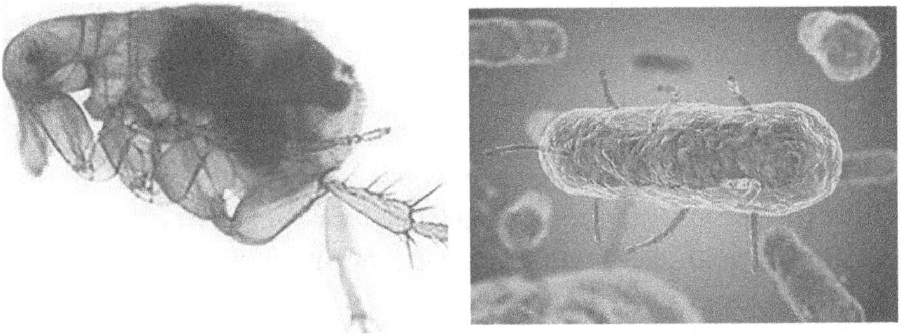

Figure 6.9 (left) Oriental rat flea (*Xenopsylla cheopis*) infected with *Y. pestis*, which appears as a dark mass in the gut; (right) SEM image of *Y. pestis.*

sodium hypochlorite [25] and 0.25 mg/L chlorine dioxide [32] are used.

Some of the most disastrous plagues recorded in history are: (i) sixth-century Justinian Plague, (ii) 1347–1353 bubonic plague (the Black Death) epidemic in Europe that accounted for the death of one-third of its population, (iii) late nineteenth-century bubonic plague epidemic in China, which was spread by rats on steamboats and killed about one million people [33–35].

Being highly pathogenic, many strains of *Y. pestis* are resistant to antibiotics and easily cultured; and therefore pose a serious threat of being used for bioterrorism purposes, especially the drug-resistant strains [36]. It is easily spread by aerosol containing plague bacilli. Jensen *et al.* [24] have reported that *Y. pestis* may have been weaponized in aerosol form and should be considered a threat, especially since its cultures were used by the Japanese Army during World War II to contaminate food and water supplies of Chinese cities [37].

6.2.1.2 Fungus

Fungi can be pathogenic and directly cause disease or it can indirectly cause disease through mycotoxins in humans, animals and plants. Hence, fungi are used as weapons against all three of them. Fungi and mycotoxins can enter the bodies of humans and animals through inhalation, ingestion, and contact with skin and mucous membranes.

Biological warfare or germ warfare uses biological toxins from fungi and infectious fungi to kill or incapacitate humans, animals or plants. A fungus known as *Cryptococcus gattii* which infects people with low immunity (e.g., AIDS or transplant patients) has been genetically manufactured to infect healthy people also. This has become a cause for concern, because if undiagnosed, this fungus enters spinal fluid and the central nervous system and causes fatal meningitis. Such research outcomes give impetus to the idea of using fungi as bioagent. The disease can spread through airborne spores. Since it has a diversity of hosts it can be used for biological warfare [38].

With the exception of *Coccidioides* spp., human pathogenic fungi are not found among lists of microbes with potential for biological warfare and bioterrorism against humans.

6.2.1.2.1 *Coccidioides* spp.

Coccidioides spp. is a dimorphic ascomycetes fungus and, in its saprobic phase, grows as a mycelium, forming a large amount of arthroconidia (Figure 6.10) in semi-arid regions with sandy, alkaline soil, high salinity and extreme temperatures. Two asexual reproductive structures are: (i) Arthroconidia that grow in the environment and are infectious to humans/animals, and (ii) Spherules with endospores that grow in the body. Known species of *Coccidioides* are *C. esteriformis*, *C. histosporocellularis*, *C. immitis*, *C. posadasii* and *C. rosea.*

Coccidioides immitis and *Coccidioides posadasii* are pathogenic species that cause fatal coccidioidomycosis (also known as Valley fever, San Joaquin Valley fever, desert rheumatism, Posadas-Wernicke disease, and coccidioidal granuloma), an infectious fungal disease caused by inhaling the spores of fungi disseminated in their natural habitat. About 60% of

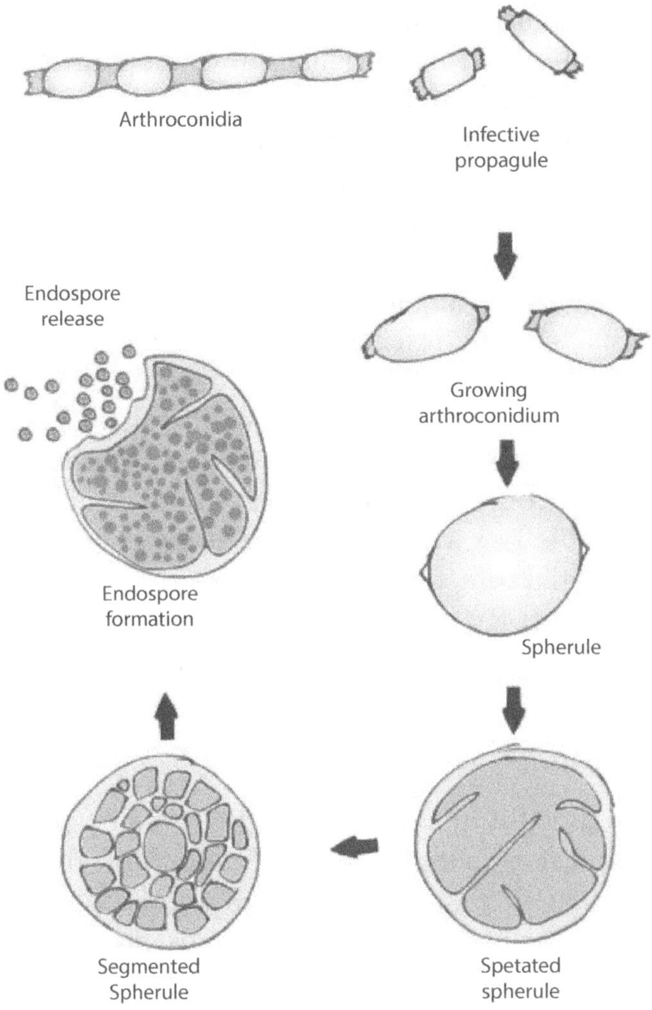

Figure 6.10 Various stages of growth of *Coccidioides* spp.

Coccidioides infections are asymptomatic. However, in the rest of patients, common clinical syndrome is an acute respiratory illness, fever, cough, and pleuritic pain, erythema nodosum (skin lesions), sometimes a severe and difficult-to-treat meningitis in AIDS and other immuno-compromised patients; and occasionally it can cause acute respiratory distress syndrome and fatal multilobar pneumonia. Tissues affected could be skin, regional lymph nodes, bones, joints, visceral organs and testes. Treatment option in humans is antifungal drugs or surgical excision. Apart from humans, this infection is common in dogs, less common in horses and cats, and rare in cattle, sheep, pigs, etc. It is difficult to prevent in endemic areas since no vaccine is available. However, it can be disinfected by using iodine, hypochlorite/bleach, phenolics, quaternary ammonium compounds, or moist heating for 15 min at 121 °C.

There is a risk of infection from dust exposure during weather-related peaks such as wet followed by dry, windy periods, and after earthquakes and windstorms. Especially at risk are farmers and construction workers. Direct transmission between people or animals is very rare. Usually the incubation period is 1–3 weeks; however, it may occur months to years after initial infection.

The World War II prisoners of war that the Japanese military held in internment camps had this infection.

6.2.1.3 Virus

Viruses are thought of as being in a gray area between a living organism and non-living chemical. Some think that a virus is an organic structure that enters, infects and interacts with all forms of living beings (i.e., human, animals, plant, microbe and archaea) [39]; whereas many accept that it is a form of life since it possesses genes that evolved by natural selection [40], and reproduce by creating multiple copies of themselves through self-assembly. However, viruses do not have a cellular structure or their own metabolism and they depend on a host cell to make new products and cannot naturally reproduce outside a host cell [41]. Outside an infected host, viruses exist as very small independent particles known as virions, having DNA or RNA, a protein coat called capsids and an envelope of lipids that surrounds capsids, which is derived from the host cell. Some of the potentiak biological warfare viruses are shown in Figure 6.11.

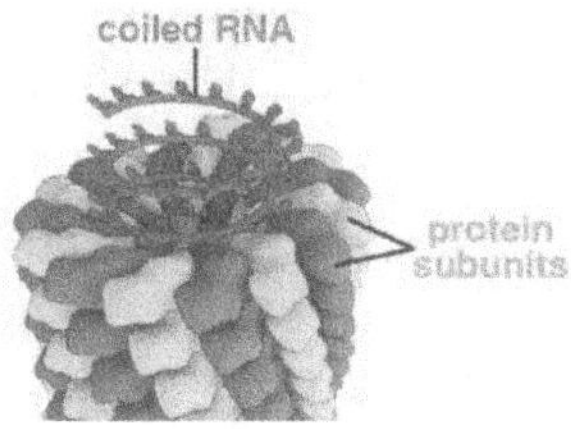

Schematic of Tobacco Mosaic Virus

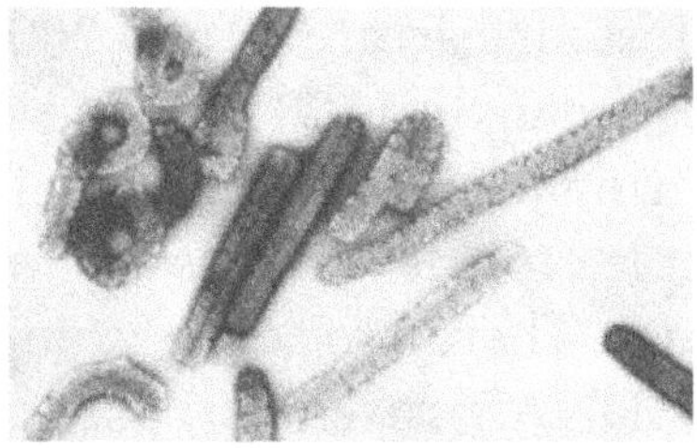

Marburg Virus (Hemorrhagic Virus)

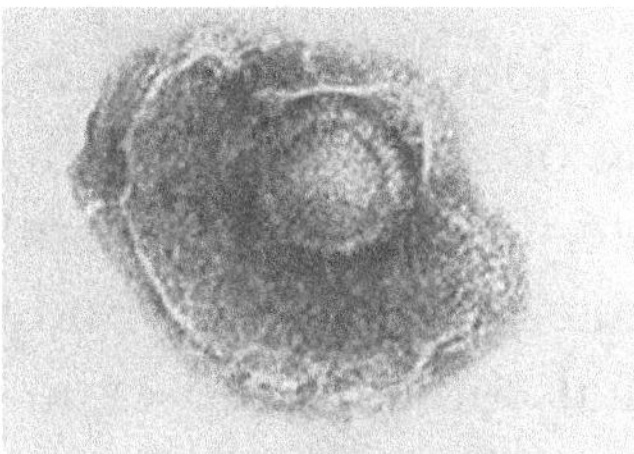

Chickenpox Virus with lipid envelope

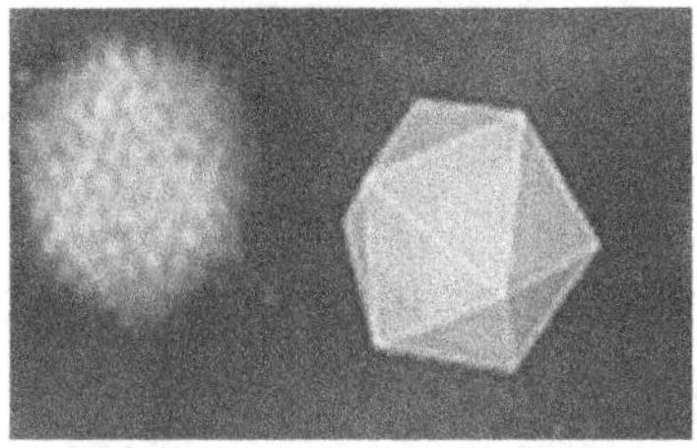

SEM image and schematic of Icosahedral Adenovirus

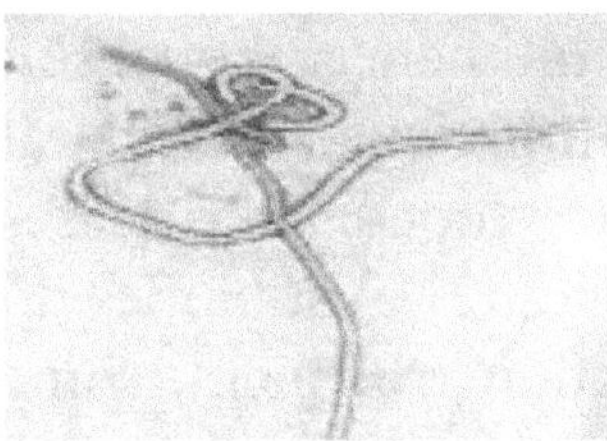

Ebola Virus (Zaire Ebolavirus)

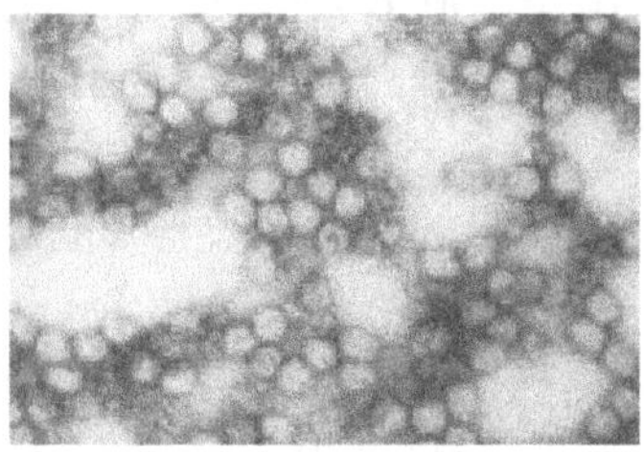

Yellow Fever Virus

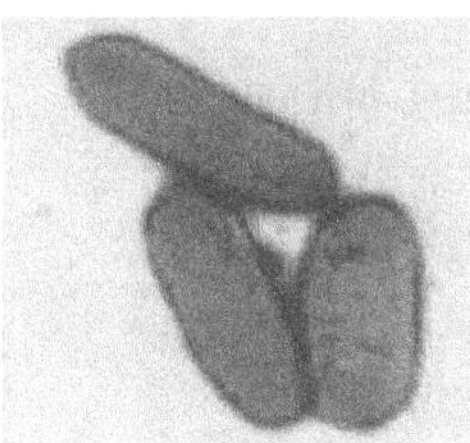

Variola Virus

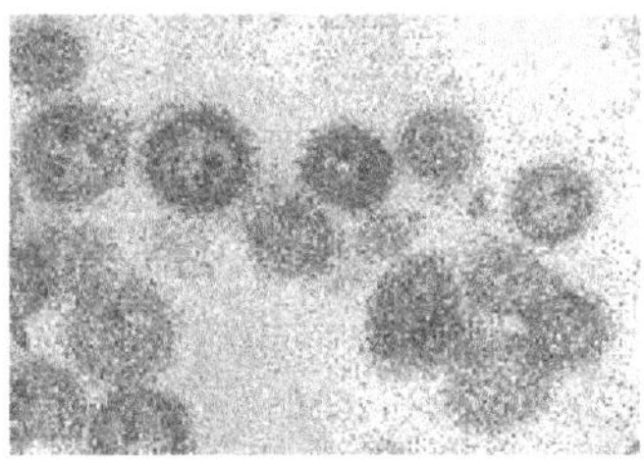

Machupo Virus

Figure 6.11 Some potential biological warfare viruses.

Most viruses are 20 and 300 nm in diameter, but some filoviruses are up to 1400 nm long and 80 nm in diameter. The shape of virions varies from helical to icosahedral, prolate or more complex forms. Viruses that infect bacteria are called bacteriophages. There are seven groups of viruses: (i) dsDNA viruses, (ii) ssDNA viruses, (iii) dsRNA viruses, (iv) (+)ssRNA viruses, (v) (−)ssRNA viruses, (iv) ssRNA-RT viruses, and (vii) dsDNA-RT viruses.

In animals, viruses are transmitted by (i) blood-sucking insects (vectors), (ii) influenza virus by coughing or sneezing, and (iii) norovirus and rotavirus of viral gastroenteritis by the fecal–oral route and from person to person by contact, entering the body in food or water, (iv) HIV through sexual contacts and exposure to infected blood. Plant viruses are transmitted by plant sap-sucking insects (aphids). Usually viral infection stimulates immune response in the host (like a vaccine does), but viruses that cause AIDS and viral hepatitis does not; hence, they result in chronic infection. To protect humans and animals from viral infection, thirteen vaccines have been developed for diseases such as polio, measles, mumps and rubella [42]. Antibiotics do not work against viruses. For viral infection treatment, some anti-viral drugs are used, such as

- Aciclovir for Herpes;
- Aciclovir, Valacyclovir, Famciclovir, and Ribavirin combined with interferon for Shingles;
- Lamivudine, TDF (tenofovir) either 3TC (lamivudine) or FTC (emtricitabine) and EFV (efavirenz) for HIV;
- Lamivudine for Hepatitis B;
- Oseltamivir (Tamiflu) and zanamivir (Relenza) for Swine Flu. [Swine flu is resistant to amantadine (Symmetrel) and rimantadine (Flumadine)];
- Oseltamivir, zanamivir, or peramivir (Rapivab) for Pneumonia;

- Brincidofovir for Ebola Virus; and
- Oseltamivir, zanamivir and peramivir (Rapivab) for Influenza.

Easy distribution of viral agents in aerosolized form has made them a potential biological warfare material. They were used by the British army in 1763 against the Native Americans during the French and Indian War. The Soviet government had planned on using the smallpox virus as a biological weapon, but it did not materialize because due to the extensive global immunization program of WHO, smallpox was eradicated. However, there are still many varieties of genetically or traditionally modified viruses that could be used as biological weapons.

6.2.1.3.1 *Phlebovirus (Bunyaviridae)* – Rift Valley Fever Virus

Rift Valley fever (RVF) is caused by *Phlebovirus (Bunyaviridae),* which is an arbovirus that affects both human and animal health. This virus has three-segmented genomes of negative-sense single-stranded RNA in viral nucleocaspid protein and enveloped by a lipid bilayer containing two viral glycoproteins, Gn and Gc [43]. This virus has not yet been fully understood [44]. No effective vaccine is available for protecting humans and livestock against this virus. Mosquitoes *(Aedes, Anopheles, Culex, Eretmapoites, Mansonia)* and sand flies are the vectors to transmit RVFV to animals through bites. Humans get infected by consuming infected meat, milk and body fluids of infected animals. In humans the symptoms of RVF are sudden onset of high fever, severe headache, myalgia, arthralgia, and photophobia. It causes necrotic hepatitis, abortion in cattle, sheep and goats, and results in about 100% mortality in young animals [45].

Due to its ability to be transmitted through aerosol, Rift Valley fever virus (RVFV) could also be used as an agent for biological warfare or bioterrorism. For example, the U.S. biological weapons program weaponized RVFV prior to termination of the effort in 1969 [43].

6.2.1.3.2 Ebola Virus (Zaire Ebolavirus)

The Ebola virus carries a negative-sense RNA genome in virions that are tubular and contain viral envelope, matrix, and nucleocapsid components. At the center of the virion structure is the nucleocapsid, around which RNA is helically wound. Tubules are approximately 80 nm in diameter, 800–1000 in length, and have a virally encoded glycoprotein (GP) projecting as 7–10 nm long spikes from its lipid bilayer surface. This virion enters the host cell by inserting the GP spike into the membrane, inducing bud formation from membrane, which forms the outer viral envelop [46].

The Ebola virus is a deadly virus that causes fatal hemorrhagic fever (severe bleeding, organ failure, and can lead to death) in humans and other mammals. It has up to 90% mortality rate [26]. Bats are the natural hosts of the Ebola virus [47]. It is transmitted in humans through body fluids [48]. The virus can be inactivated by exposing it to 56 °C for 30 min, by 1–2% sodium hypochlorite and/or 1% iodine. Suitable anti-Ebola drugs include Brincidofovir, Favipiravir, Lamivudine, Triazavirin, BCX4430, DZNep, FGI-103, FGI-104, FGI-106, JK-05 and ZMapp. Effective vaccines are cAd3-ZEBOV and rVSV-ZEBOV.

Because of its high mortality rate, it may have been weaponized for aerosol application.

6.2.1.3.3 *Flaviviridae* (especially *Japanese Encephalitis Virus*)

The family *Flaviviridae* includes three genera: *Flavivirus* (yellow fever virus), *Hepacivirus* (hepatitis C virus) and *Pestivirus* (bovine viral diarrhea virus). *Flavivirus and Hepacivirus* viruses are associated with human disease and mosquitoes are the vector. *Flaviviridae* are spherical in shape, 40 to 60 nm in diameter, and contain a single positive-strand RNA genome surrounded by an icosahedral nucleocapsid covered in a lipid bilayered envelope [49]. Like the Ebola virus, in this virus the envelope is also derived from the host cell membrane.

Apart from Japanese encephalitis virus (JEV), this species is also related to Murray Valley encephalitis virus (MVEV), St. Louis encephalitis virus (SLEV), and the West Nile virus (WNV). JEV spreads from a mosquito (Culex species) that has sucked the blood of an infected human and its incubation period is 515 days. Initial symptoms are fever, headache, nausea, vomiting, and drowsiness, which eventually leads to encephalitis (inflammation of brain due to infection). In the case of severe infection, permanent neurological or psychiatric damage may occur. Formalin-inactivated vaccine and a live-attenuated vaccine are used for JEV. No anti-viral drug has been developed against the yellow fever virus. Use of JEV has not been ruled out as a biological warfare agent.

6.2.1.3.4 Machupo Virus

The Machupo virus is an arenavirus endemic to Bolivia, it is also known as viral hemorrhagic fever (VHF), Bolivian hemorrhagic fever (BHF), South American hemorrhagic fever, black typhus, or dog fever. Arenaviruses have a single-stranded, bi-segmented RNA genome; a large (7200 nt) and a small (3500 nt) segment and a lipid envelope with 8–10 nm club-shaped projections.

The Machupo virus causes hemorrhagic fever. It has an incubation period of 7–16 days with slow onset of symptoms. The first symptoms are fever, malaise, headache, muscular pain, anorexia, nausea and vomiting. Then, between the third and fifth day, dehydration, hypotension, infrequent urination, and bradycardia occurs, leading to the hemorrhagic phase, which starts from the nose, gums, stomach and intestines where severe blood loss results in hypotensive shock and neurological crisis. For Machupo therapy, Ribavirin and immune plasma therapy are suitable.

In the conclusion of their article "Hemorrhagic Fever Viruses as Biological Weapons," Groseth *et al.* [50] have written that "In addition to causing illness and death, a biological attack would aim to cause fear in the general populace and, thus, result in social and economic disruption. Based on their fearsome reputation and dramatization by the popular media, VHF agents would be excellent candidates to serve this purpose. Given the potential for a biological attack to occur, it is of the utmost importance that resources and knowledge are made available to deal effectively with such a situation in a safe and timely manner."

6.2.1.3.5 *Marburg marburgvirus* (Marburg Virus)

The *Marburg marburgvirus* is another hemorrhagic fever virus. This virus belongs to the same family as the Ebola virus and is considered to be extremely dangerous as it has a fatality rate of 50–88%. Marburgvirions are filamentous particles that may be shaped as a shepherd's crook, "U" or "6"; or coiled, toroid or branched [51]. They are approximately 80 nm in width, but their length varies from 795 to 828 nm. Marburgvirions consist of genomic RNA wrapped around a polymer of nucleoproteins (NP) at the center. They have seven structural proteins that are surrounded by a lipid membrane derived from the host cell membrane. The membrane anchors a glycoprotein ($GP_{1,2}$) that projects 7 to 10 nm spikes away from its surface.

Fruit bats (*Rousettus aegyptiacus*) are the natural hosts of Marburg virus. The Marburg virus is transmitted to people from fruit bats, especially those who have prolonged exposure to fruit bat inhabited caves or mines. It spreads among humans through human-to-human transmission via direct contact (through broken skin or mucous membranes) with the blood, secretions, organs or other bodily fluids of infected people, and with surfaces and materials (e.g., bedding, clothing) contaminated with these fluids and via infected semen. The treatment needs oral or injected rehydration, a range of blood products, immune therapies and drug therapies are currently under development.

If pregnant women are infected it persists in their breast milk and in the fetal placenta.

The symptoms of Marburg virus (MARV) disease appear in 2–21 days with abrupt high fever, severe headache and malaise, muscle aches, watery diarrhea, abdominal pain and cramping, nausea and vomiting. Hemorrhage in multiple areas (in vomit, feces, from the nose, gums and vagina) sets in between 5–7 days. Patients often show "ghostlike" drawn features, deep-set eyes, expressionless faces, and extreme lethargy. In the 1967 European outbreak, a non-itchy rash was a feature noted in most patients between 2 and 7 days after onset of symptoms. The central nervous system (CNS) gets affected, resulting in confusion, irritability, and aggression; and, at a later stage, orchitis (inflammation of one or both testicles) also may occur, leading to death. In fatal cases, death occurs most often between 8 and 9 days after symptom onset, usually preceded by severe blood loss and shock.

In his book, Soviet defector Steven Alibek wrote that Marburg virus was included in the biological weapons research program of the USSR during the Cold War, which killed a researcher. He claimed that a weapon filled with MARV was tested at the Stepnogorsk Scientific Experimental and Production Base in the Kazakh Soviet Socialist Republic (today Kazakhstan) [52].

6.2.1.3.6 Variola Virus

Variola major is a virus that causes smallpox. Smallpox used to be a contagious, disfiguring, and often deadly (fatality rate of 25%) disease, but is now globally eradicated (last known naturally occurring case in the world was in Somalia in 1977) except in laboratory stockpiles. However, it cannot be overlooked as a biological warfare concern, as it was explored by the Japanese Army for weaponization during World War II [26, 29], who considered it suitable for dissemination through contaminated items.

Variola virus is a rather large (300×200 nm in size), rectangular, brick-shaped virus having an envelope derived from a host cell. The core is of double-stranded DNA. It is enveloped and contains filamentous nucleocapsid. It spreads by prolonged person-to-person contact or via inhaled droplets of air or aerosols that enter through the respiratory mucosal route. It has an incubation period of 12 days; symptoms after infection are chills, fever, prostration, headache, backache, vomiting as well as pustule formation. The variola virus has been reported to survive in a scab for 13 years [53] and it is resistant to drying [25]. Variola virus is inactivated by 1% sodium hypochlorite. The vaccine that eradicated smallpox provided a high level of protection against the disease for about five years. Without the vaccine, it is nearly impossible to prevent infection. The vaccine contains live virus; it does not contain smallpox and does not cause smallpox.

Smallpox was used as a biological weapon during the French and Indian War in the United States (1754–1767), when British soldiers gave the Indians blankets that had been used by smallpox patients; and, as previously mentioned, Japan considered the use of smallpox as a biological weapon in World War II. Smallpox is currently contained in two high security labs in the United States and Russia, and this virus is rated among the most dangerous of all potential biological weapons.

6.2.1.3.7 Yellow Fever Virus

Yellow fever virus is a member of the *Flaviviridae* family. It was the first disease known to be caused by a virus, the first isolated virus and the first reported with an insect as its vector. This virus has positive-sense, single-stranded RNA of 11,000 nucleotides long and a single polyprotein. Yellow fever is a hemorrhagic fever, which spreads by the bite of an infected female mosquito (*Aedes aegypti*) (Figure 6.12). The symptoms of this viral fever are chills, fever, loss of appetite, nausea, headache, backache and muscle pains. Symptoms typically improve within five days, but in some cases the fever comes back with

Figure 6.12 *Aedes aegypti* (Source: James Gathany – CDC-PHIL, Public Domain, https://commons.wikimedia.org/w/index.php?curid=54701000).

abdominal pain and the liver becomes damaged, causing yellow skin (hence the name yellow fever), which leads to bleeding risks and kidney problems.

Yellow fever can be prevented by vaccination and avoidance of mosquito bites. No drug is available to cure yellow fever. However, medical treatment focuses on easing symptoms such as dehydration; and drugs, such as paracetamol (acetaminophen), are given for pain relief.

In his book entitled Yellow Fever: A Deadly Disease Poised to Kill Again, while writing about weapons of mass destruction, Dickerson has hinted that the yellow fever virus could be used as a biological weapon [54].

6.2.1.3.8 Venezuelan Equine Encephalitis, Eastern Equine Encephalomyelitis and Western Equine Encephalomyelitis Viruses

Viruses from family *Togaviridae* and genus *Alphavirus* have three different species, each causing different equine encephalomyelitis, namely, Venezuelan equine encephalitis virus (VEEV), Eastern equine encephalomyelitis virus (EEEV) and Western equine encephalomyelitis (WEEV). As the name suggests, they infect equine species, such as horses, donkeys, and zebras, and affect their central nervous system and the animal suddenly dies. Humans can also contract this disease. WEE is uncommon in humans; however, the virus can infect infants and children.

The VEEV, EEEV and WEEV viruses are all arboviruses (arthropod-borne virus) transmitted to humans by mosquitoes that have bitten infected animals.

The VEE virion is spherical, approximately 70 nm in diameter, and has a lipid membrane with glycoprotein surface proteins spread around the outside. Surrounding the nuclear material is a nucleocapsid that has an icosahedral symmetry of T = 4, and is approximately 40 nm in diameter.

There is no drug or vaccine for WEE.

During the Cold War, both the United States biological weapons program and the Soviet biological weapons program researched and weaponized VEEV [55]. According to a report by Middlebury College, EEEV and WEEV were one of more than a dozen agents that the United States researched as potential biological weapons before the nation suspended its biological weapons program.

6.2.1.4 Insects

Use of insects or entomological warfare (EW) has also been used to attack the enemy. It is a part of biological warfare that has existed for centuries and also is being researched in the modern era [68]. Japan and several other nations have developed and been accused of using an entomological warfare program. In EW, insects, such as bees, wasps, etc., are employed either in a direct attack or as vectors to deliver a biological agent such as plague. EW involves infecting insects with a pathogen and then dispersing the insects over targeted areas against humans, animals, agricultural crops, fisheries or aquatic plants. For agricultural crops, the insects may not be infected with any pathogen but instead represent a threat to crops by feeding on them or using them in the breeding process. Insects used:

- ***Harlequin Bugs (Murgantia histrionic)*** destroy cabbage; they were used during the American Civil War.
- ***Colorado Potato Beetles*** affect potatoes; they are resistant to dichlorodiphenyltrichloroethane (DDT) and other major pesticide classes [56]; they were used by the Germans during World War II to destroy enemy food sources [10].
- ***Oriental Rat Fleas*** were used as a vector of murine typhus and bubonic plague, which are diseases transmitted from one flea generation to the next by their eggs [57]. Japan used them against the Chinese in World War II.
- ***House Flies*** are a vector of up to 100 pathogens that can cause cholera, typhoid, salmonellosis, anthrax, tuberculosis, ophthalmic diseases, bacillary dysentery, parasitic worms and viruses like poliomyelitis, enteroviruses and viral hepatitis [58–60]. Some strains are resistant to many kinds of commonly used pesticides [61, 62]. The Japanese also used cholera infected flies as an entomological weapon against the Chinese to spread disease during World War II [10].
- ***Ticks*** can be used as a biological weapon to spread different diseases in poultry, livestock and humans. They are carriers of many bacteria, protozoans and viruses [63]. The fowl tick *Argas* spp. causes aegyptianellosis in poultry [64]. Ticks are a vector of *Babesia bigemina*, which causes cattle fever. In birds (chickens), ticks transmit *Chlamydophila psittaci*. In humans, they are responsible for causing rickettsialpox, typhus, African tick bite fever, boutonneuse fever, Rocky Mountain spotted fever, Queensland tick typhus, Flinders Island spotted fever, Colorado tick fever, Q fever, tick-borne meningoencephalitis and ehrlichiosis [65]. During the Cold War era,

the Soviet Union developed techniques to transmit foot-and-mouth disease using ticks.

- ***Mosquitoes*** can be used as a biological weapon to cause disease in birds, animals and humans. They are the vectors of different viruses and parasites. Viral diseases include dengue fever, chikungunya and yellow fever, which are mostly caused by *Aedes aegypti* [66]. Similarly, malaria is caused by the protozoan parasite *Plasmodium* [67]. During the Cold War, the United States made a laboratory which can produce 100 million yellow fever infected mosquitoes to attack the Soviet Union [10]. In 1955, the United States dropped 300,000 yellow fever infected mosquitoes during Operation Big Buzz to check their survivability [63].
- ***Caterpillars*** are the larval stage of members of the order Lepidoptera, the insect order comprising butterflies, moths, sawflies, etc., which are herbivorous and agricultural pests [69] which eat leaves. Many species are resistant to many pesticides [70]. In the 1990s the United States researched the use of caterpillars as a biological weapon against crops [71].
- ***Black Flies*** are small (3–6 mm in size), robust flies with short wings and humped thorax. They feed on the blood of poultry and cattle, which could be fatal due to acute toxemia. They are also vectors of protozoans (*Leucocytozoon*), which cause leucocytozoonosis in poultry, and filarial nematodes (*Onchocerca*), which cause bovine onchocerciasis [63]. In humans, black flies transmit onchocerciasis, which are vectors of the parasitic nematode *Onchocerca volvulus* that lives on human skin and is transmitted to humans by the bite (blood meal) of the fly; therefore, they can also be used as a

biological weapon to transmit disease to humans, poultry and cattle [72].

- ***Biting Midges / Punkies / Sand Flies*** are 1–4 mm long flies. They are vectors of arboviruses [73] and some non-viral pathogens [74]. The viruses cause bluetongue in cattle and sheep and spread blood protozoans in poultry and equine onchocerciasis in horses. It is believed that artificially infected biting midge colonies can be used to spread disease among cattle and poultry.
- ***Horse Flies and Deer Flies*** are pests that transmit different diseases in cattle, causing deep bleeding wounds [75]. They are robust, large flies with blade-like mouth parts that inflict the wounds. They transmit bovine leukemia, infectious anemia, hog cholera, *Trypanosoma* sp. protozoans and *Elaeophora* sp. nematodes [76]. They can also be used as biological weapons.
- ***Warble Flies*** are large parasites of cattle. Their larvae are often called "wolves" or "cattle grubs." They are large, hairy, orange to yellow in color and bee-like in appearance. The adults are free living and have vestigial mouthparts [77]. Larvae of some species invade human tissues. Warble fly larvae can be used as a biological weapon because in humans the larvae of *Hypoderma bovis* is known to cause intracerebral myiasis, in which the invasion of intracerebral tissues takes place, causing intracerebral hematoma and convulsions [78]. *Hypoderma tarandi* is parasitic on reindeer/caribou and may also cause ophthalmomyiasis in humans, leading to glaucoma, uveitis, and retinal detachment [79].
- ***Screwworm Flies*** are parasitic flies whose larvae infest open wounds of animals [80] and attack

healthy living tissues of their host. It causes tissue loss, vital organ destruction and even death in extreme cases. Since female screwworms can lay up to 3000 eggs in her lifecycle and travel up to 200 km in search of a host [81], they can also be used as a biological weapon against livestock.

6.2.1.5 *Biogenic Toxins*

According to *Dictionary.com,* the definition of a toxin is "an antigenic poison or venom of plant or animal origin, especially one produced by or derived from microorganisms and causing disease when present at low concentration in the body."

Through their chemical effect on living processes, biotoxins cause death, temporary incapacitation or permanent injury in people and animals. There are nine biotoxins that have been or are considered for weaponization and should be considered threats. Toxins are delivered through aerosols; hence, they are initially a threat to the respiratory tract. According to Franz [82], the effects of toxins are more severe when inhaled than when consumed in food. Some biotoxins are expected to be threats to drinking water also, and some toxins can elicit a significantly different clinical picture when the route of exposure is altered, which may confound the diagnosis and treatment. The nine biotoxins are briefly described below.

6.2.1.5.1 Anatoxin-a

Anatoxin-a ($C_{10}H_{15}NO$) is a secondary, bicyclic amine alkaloid and cyanotoxin produced by filamentous cyanobacteria *Anabaena flos-aquae* found in freshwater. It also is produced by other cyanobacteria such as *Aphanizomenon, Cylindrospermum, Microcystis, Oscillatoria, Planktothrix,* and *Raphidiopsis.* It is unstable in water and under ambient conditions in water and is converted to nontoxic form. It cannot be degraded by alum flocculation, filtration, and chlorination. Vuori *et al.* found that a water purifier containing carbon ion exchange resin and silver could reduce only 50% anatoxin-a concentration [83]. Reverse osmosis (RO) is completely effective.

Anatoxin-a is a neurotoxic material, which upon exposure causes very fast death by respiratory paralysis.

6.2.1.5.2 Botulinum Toxin

Clostridium botulinum produces botulinum ($C_{6760}H_{10447}N_{1743}O_{2010}S_{32}$), a powerful nerve toxin; about 1 μg is lethal to humans. There are eight types of botulinum, i.e., A–H. Botulinum A, B, E and F cause disease in humans, with H being the deadliest. However, A and B are used as medicine, e.g., Botox. Botulinum toxin can be absorbed through the eyes, mucous membranes, respiratory tract, or non-intact skin. It blocks nerve function and leads to respiratory and progressive head-to-toe musculoskeletal paralysis, where victims remain mentally alert until death due to the inability to breathe. Botulinum toxins are more toxic when ingested than when inhaled [28].

Botulinum toxin is inactivated by Sunlight in 1–3 h, air can detoxify it within 12 hours, and heating at 80 °C destroys it in 30 min. It can be detoxified by boiling or sterilization [84]. Treatment using charcoal is also recommended to effectively remove the toxin [29].

Botulinum toxin has been recognized as a potential agent for use in bioterrorism. It has been weaponized by Iraq and others for aerosol application [26, 28, 85, 86].

During the 1990s, the Japanese doomsday cult Aum Shinrikyo spread botulinum toxin as an aerosol, but the attacks did not cause any fatalities. The Red Army Faction, a German terrorist group, also produced botulinum toxin, but never used it in any attack.

6.2.1.5.3 Enterotoxin B

Enterotoxin B is a protein produced by *Staphylococccus aureus;* it is also known as Staphylococcal enterotoxin B (SEB). It is an incapacitating toxin. If ingested it causes severe gastrointestinal

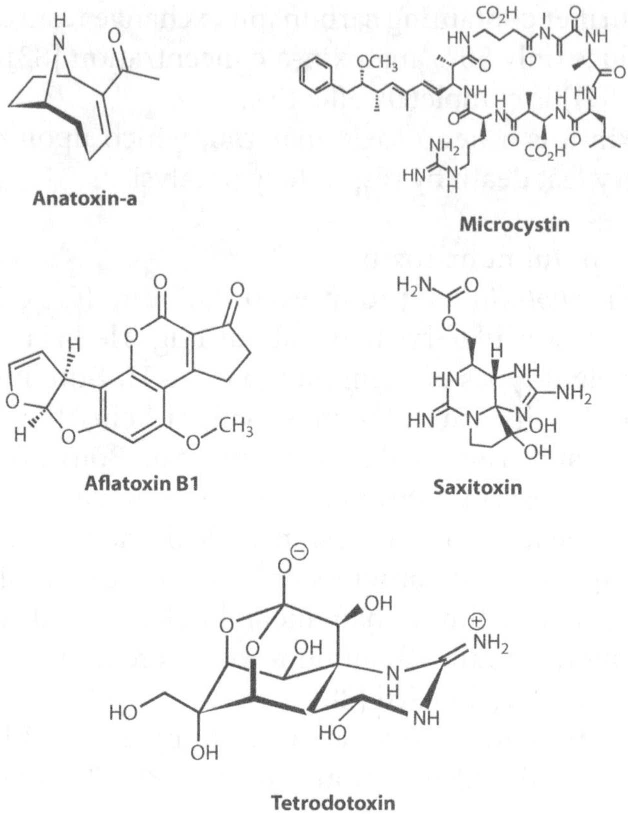

Figure 6.13 Molecular structures of some toxins considered as potential biological warfare agents. *Anatoxin-a* (Source: Cacycle - Own work, Public Domain https://commons.wikimedia.org/w/index.php?curid=3204512). *Microcystin* (Source: Wikipedia). *Aflatoxin B1* (Source: Wikipedia. https://commons.wikimedia.org/w/index.php?curid=10810336">Link</a>). *Saxitoxin* (Source: Wikipedia. https://commons.wikimedia.org/w/index.php?curid=50713663). *Tetrodotoxin* (Source: Wikipedia).

pain, projectile vomiting and diarrhea. If inhaled it causes fever, chills, headache, muscle aches, shortness of breath, and nonproductive cough [23, 26, 82]. Signs and symptoms develop within several hours. Several other microbes contain enterotoxin to create similar effects as SEB, e.g., *E. coli*. SEB can be either inhaled or ingested from contaminated water or food.

Since SEB has been weaponized, it could be used to sabotage food or low volume water supplies. Full recovery is likely, but soldiers could be incapacitated for up to two weeks [26]. Ulrich *et al.* [87] have stated that 1.7 pg/person is lethal by aerosol. SEB is stable in both acidic and basic solutions but does not survive for long at room temperature [25]. It is inactivated at 100 °C. No disinfecting efficacy of SEB is known.

6.2.1.5.4 Epsilon from *Clostridium perfringens*

Epsilon is a proteinaceous toxin produced by *Clostridium perfringens,* a Gram-positive, rod-shaped, anaerobic, spore-forming bacteria, which also produces two other toxins, α and β. There are five strains of *C. perfringens* A–E. Only the B and D strains produce epsilon. *C. perfringens* type B causes severe enteritis in young calves, foals, lambs and piglets; and type D causes enterotoxaemia in sheep, goats and cattle. Both α and β toxins are also potential biological weapons. The *α*-toxin is produced by all strains and is lethal by aerosol and is a necrotizing toxin that causes serious acute pulmonary disease, as well as vascular leak, hemolysis, thrombocytopenia and liver damage. *β*-toxin is also a lethal necrotizing toxin found in type B and C strains and it can damage blood vessels, resulting in leukostasis, thrombosis, decreased perfusion and tissue hypoxia. In humans, the disease is usually caused by type A and C strains, by food poisoning. It has been isolated from human wound. Wound contamination causes gas gangrene, clostridial cellulitis or superficial contamination.

The epsilon toxin is transmitted through contaminated food, water or by aerosol. Being a protein, it can be denatured thermally. In natural infections with type B strains, hyperimmune serum and antibiotics have been found helpful. Toxoid vaccines can prevent type B and D enterotoxaemia.

6.2.1.5.5 Microcystin-LR

Microcystin-LR contains many non-proteinogenic amino acids such as dehydroalanine derivatives and *β*-amino acid (ADDA).

It is produced by the freshwater cyanobacteria (blue-green algae) *Microcystis aeruginosa, Anabena, Oscillatoria, Nostoc,* etc. that grows in water rich in phosphorus. Microcystins are liver toxins and tumor-promoting toxins. Microcystin causes pansteatitis (yellow fat disease, which is a physiological condition in which the body fat becomes inflamed). When ingested, the microcystin goes to the liver and causes severe liver damage. Mycrocystins do accumulate in fish.

Microcystins are stable in a wide range of pH and are soluble in water. Most of them cannot penetrate directly through plant or animal cell membranes. Uptake into cells occurs through membrane transporters. The liver is the ultimate target organ for toxic effects. Microcystins are lethal at low doses. They persist in dark water for months or even years. Bacteria and sunlight break down microcystin. Plastic containers other than polyethylene terephthalate (PETG) can absorb microcystin.

6.2.1.5.6 Mycotoxins: Aflatoxins

Aflatoxins are metabolites of the fungi *Aspergillus flavus* and *Aspergillus parasiticus,* which infect a variety of agricultural plants, including peanuts, cereal grains, tree nuts, oil seeds, spices, etc. They are difuranocoumarins. The water solubility of aflatoxins is limited and they are heat stable.

Aflatoxins are potent mutagens and carcinogens. However, they are less toxic than botulinum, staphylococcal enterotoxin or ricin. There are fourteen different natural aflatoxins.

Exposure to aflatoxins occurs through food and milk; and occupational exposure, such as working in oil mills, also occurs. Aflatoxin poisoning causes jaundice, rapid development of ascites (abdominal fluid) and portal hypertension, and gastrointestinal bleeding leading to mortality. In children, aflatoxin causes Reye's syndrome, leading to disturbed consciousness, fever, convulsions and vomiting [88]. Aflatoxin B is a strong carcinogen.

Aflatoxins have been weaponized by Iraq for missile delivery [85, 86].

6.2.1.5.7 Ricin

Ricin is a plant metabolite produced by the beans of *Ricinus communis* (castor beans). Castor oil waste contains 3–5% ricin [82] protein. It is a ribosome inactivating globular, glycosylated heterodimer protein.

Ricin is not a threat as a weapon of mass destruction because it is required in large quantities to affect mass populations. However, it has been used as an assassin's weapon. Simon [89] has reported that an umbrella was used for injecting a "ricin ball" into a human target. If injected, it quickly affects the central nervous system, decreases heart function and convulsions occur after the toxin blocks synapse actions in the brain, finally resulting in death.

If ingested, ricin causes gastrointestinal hemorrhage with organ necrosis [26]. Ricin is detoxified by heating at 80 °C for 10 minutes and for 1 hour at 50 °C [25]. It is stable under ambient conditions [84]. Reverse osmosis (RO) removes > 99.8% ricin from water.

Ricin is a potential aerosol threat. It is in the biological warfare inventory of many countries. The United States investigated ricin for its military potential during World War I as a toxic dust or coating for bullets and shrapnel, but the war ended before the United States weaponized it. During World War II the United States and Canada undertook studying ricin in cluster bombs [90], but it was found that it was not more economical than using phosgene; moreover, the United States diverted their attention to weaponized sarin.

The Soviet Union and Iraq [86] also possessed weaponized ricin.

6.2.1.5.8 Saxitoxin

Saxitoxin is produced by marine algae (cyanobacteria or dinoflagellates) *Gonyaulax spinifera*. It is present in bivalve shellfish such as mussels, clams, oysters and scallops. It is a neurotoxin that blocks the sodium channel of neurons. Saxitoxin is water soluble, acid stable, alkaline labile, and stable under normal atmospheric conditions [84].

Saxitoxin is highly toxic if shellfish contaminated by toxic algal blooms is ingested, injected or administered by aerosol [91]; then within 30 min of uptake, abdominal distress, diarrhea, nausea, vomiting, vertigo, headache, rapid pulse, numbness of the tongue and gums occur, which finally leads to paralysis [92] and possibly death in 1–24 h if respiratory failure sets in. The only known treatments are general supportive care and artificial respiration. Up to > 98% saxitoxin can be removed from water by RO.

It has been weaponized for covert purposes as a chemical weapon by the U.S. military [93]. Wheelis *et al.* [94] have written that the CIA issued a small dose of saxitoxin to U-2 spy plane pilot Francis Gary Powers in the form of a small injection hidden within a silver dollar, for use in the event of his capture and detainment.

6.2.1.5.9 Tetrodotoxin

Like saxitoxin, tetrodotoxin ($C_{11}H_{17}N_3O_8$) also blocks the sodium channel of neurons and is a potent lethal neurotoxin. This toxin is present in fish of Tetraodontiformes order, such as puffer fish, porcupine fish, triggerfish, etc., and also in blue-ringed octopus. It is produced by infecting or symbiotic bacteria like *Pseudoalteromonas, Pseudomonas*, and *Vibrio.* It has also been synthesized chemically [95]. Tetrodotoxin is soluble in slightly acidic water and is temperature stable [84]. Consumption of improperly prepared puffer fish has been known to cause death [91]. The toxic signs of tetrodotoxin poisoning, which appear in 10 minutes to 4 hours after ingestion, are numbness of the lips, tongue, and fingers, anxiety, nausea, vomiting, paralysis, respiratory failure, leading to death, usually within six hours [96]. The prognosis for recovery is good if artificial respiration is given and the patient survives the first 24 hours.

Tetrodotoxin is inactivated by chlorine at highly acidic (< 3) or alkaline pH (> 9) [91]. Being soluble in water, it was investigated as a potential biological warfare weapon [27] and a threat to drinking water [83].

6.3 Chemical Warfare

Chemical warfare is intentional application for hostile purposes of toxic substances against humans and their environment.

Toxic chemicals used in warfare may be in solid, liquid or gaseous form (Figure 6.14). Depending on the type of chemical used, the health effects caused by these chemicals include dizziness, nausea, blindness, disorientation, serious injury, immobilization, and even death. Immediate attention and direct mitigation measures are necessary, such as, moving away from the area where the chemical agent was released to get fresh air, putting on a filter mask, removing clothes, cleansing the skin, washing out eyes, and administering an antidote.

Chemical warfare agents are highly toxic but comparatively easy to handle as they are capable of being stored for long periods in containers without degradation and corroding. Containers have to be resistant to atmospheric water and oxygen, and should withstand the heat developed when dispersed with an explosive device against crops. Both chemical and biological agents are used for anti-agricultural (including crops, vegetation, fisheries) and anti-livestock warfare. The United States developed herbicides and chemicals for causing crop plant disease for destroying enemy agriculture, fisheries and water-based vegetation to thwart Sino-Soviet aggression. Wheat blast and rice blast were weaponized in aerial spray tanks and cluster bombs for delivery to enemy watersheds. They also carried out scorched earth tactics of destroying livestock and farmland during the Vietnam War and the Eelam War in Sri Lanka. In the 1980s, a Soviet Ministry of Agriculture and Food secret program having the code name "Ecology" successfully developed variants of foot-and-mouth disease and rinderpest against cows, African swine fever against pigs, and psittacosis against chickens.

6.3.1 Types of Chemical Weapons

The chemical agents may be (i) persistent or (ii) non-persistent:

(i) *Persistent Chemicals* — These chemicals do not evaporate or break down quickly. The breakdown of chemicals depends on rain, humidity, wind, temperature and surface type.
(ii) *Non-Persistent Chemicals* — These are chemicals which quickly break down. They are usually gases and liquids that evaporate quickly such as chlorine and ammonia.

Based on the type of harm that chemical agents cause, they can be categorized as follows.

6.3.1.1 Nerve Agents

When inhaled, within minutes nerve agents affect the nervous system, causing convulsions, paralysis and death. They are persistent from a few hours to a few weeks. Some of the nerve agents in use are:

- Sarin (GB) $C_4H_{10}FO_2P$, which was used by the Aum Shinrikyo cult in the trains of the Tokyo subway in March 1995, is an example of a non-persistent and extremely toxic nerve gas. It is a colorless, odorless organophosphorus compound. At high concentrations, it can kill within 2–15 minutes of exposure.

Examples are the G-series nerve agents tabun and soman. Symptoms are blurred vision, chest tightness, nausea, vomiting, increased heart rate, loss of consciousness, paralysis and death.
- VX (methylphosphonothioic acid), VE, VG, VM-V Series are persistent nerve gas that can kill within a few hours. Other nerve agents are GA (tabun) and GD (soman).
- Multiple chemical agents, possibly mustard gas and the nerve agents sarin, tabun and VX, and the blood agent hydrogen cyanide may have been used against Iranian troops and Iraqi Kurdish guerrillas allied with Tehran during the Iran–Iraq War.

6.3.1.2 *Blister Agents*

Blister agents are often persistent oily droplets that upon contact with skin, within minutes or hours irritate, cause blisters and destroy the skin; upon contact with eyes they can cause blindness; if inhaled they may result in fatal respiratory damage. Others in this group are HN or nitrogen mustard. Some of the common blister agents are:

- Mustard gas (2-chloroethyl sulphide) or HD, which is truly an oily liquid, which is colorless and odorless. It can be released using a balloon or airplane. It attacks blood-generating organs (spleen, born marrow and lymphatic tissues), reduces WBC and immune defenses and causes severe skin and lung injuries. Its symptoms do not appear until 2–24 hours after exposure. There are no antidotes for it.
- Lewisite or Arsine ($C_2H_2AsCl_3$) is an aliphatic arsenic compound and is a highly volatile colorless liquid that cause injuries similar to mustard gas but

much more quickly. It destroys the ability of the hemoglobin (Hgb) component of the blood to bind oxygen. Its antidote is dimercaptopropanol.

6.3.1.3 *Choking Agents*

Usually choking agents are not persistent. If inhaled, they cause blood vessels to hemorrhage and fluid to build up in the lungs until the victim chokes or drowns in their own fluids. The smell or lung irritation gives warning of the presence of these agents. Reactions or symptoms are seen in seconds or up to three hours after exposure, for example:

- Chlorine gas quickly burns tissues of the nose, mouth and lungs. Germany used 168 tons of chlorine gas against French Algerian and, later, Canadian troops.
- Carbonyl chloride (CS) or Phosgene are a lethal type of choking and irritant agents. It can act as an aerosol and penetrate the body through the respiratory system. Symptoms are irritation and eye corrosion causing blurred vision; a long-term symptom is lung fibrosis, causing impaired lung functions.
- Phosgene ($COCl_2$) is a colorless gas; it smells like freshly cut hay or grass. In 1812, during WWI, it was used by the German army. Along with diphosgene, a liquid which is equally toxic, it is a part of the chemical arsenal of many countries, including the United States.

6.3.1.4 *Blood Agents*

Blood agents are non-persistent agents which are very fatal when inhaled. They initially cause headache, nausea, and

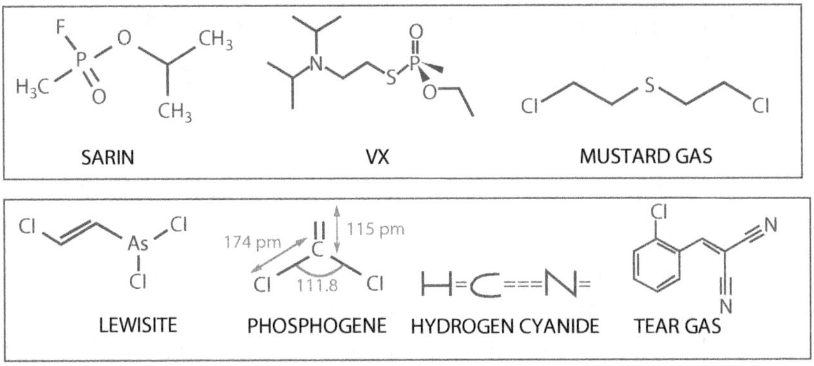

Figure 6.14 Structural formula of some chemical warfare agents.

vertigo, as they interfere with oxygen use at the cellular level. Examples of blood agents are:

- Hydrogen cyanide (HCN), cyanogen chloride and cyanide salts are very poisonous and at sufficient concentrations rapidly lead to death, e.g., 300 mg/m^3 kills immediately, whereas 200, 150, 120 and 40–60 mg/m^3 are fatal after 10, 30 and 30–60 minutes respectively. Symptoms are restlessness, giddiness, respiration difficulty, unconsciousness or sudden collapse and death. Known antidotes are dicobalt edetate and hydroxocobalamin.

6.3.1.5 Riot Control Agents

Riot control agents are also non-persistent and rarely fatal. They include:

- Pepper sprays, which are extracts from black pepper.
- Tear gas is used for riot control; 2-chlorobenzalmalononitrile ($C_{10}H_5ClN_2$), a cyanocarbon, is the defining component of tear gas, commonly referred to as CS gas.

6.4 How Nanotechnology Can Protect from Biological and Chemical Warfare

The impact of chemical and biological agents on humans, animals or plants may remain undetected for hours, days or weeks, depending on the agent, e.g., anthrax letters. The present-day scenario of defense concerns demands on R&D efforts to develop passive defense capabilities, counter proliferation, and rapidly tackle the consequences of various chemical and biological weapons. By exploiting the advantages of modern technologies, such as nanotechnology, information technology (IT), bioengineering, multifunctional materials and human performance studies, protective measures can be tackled, including

1. The ability to sense an object of concern by locating, detecting and identifying toxins and pathogenic microbes;
2. The ability to shield or protect an individual by preventive measures, medical treatments, etc.; and
3. By further decontamination, restoration, and post-exposure theranostics.

6.4.1 Nanosensors that Aid Biological and Chemical Warfare

Use of nanosensor is highly imperative in defense because nanosensors are smaller, require less power, have greater sensitivity and better specificity. Many sensors that utilize the unique properties of nanomaterials to become smaller and more sensitive compared to conventional technology have already been developed. Portable nanosensors will be highly valuable to military field operatives; especially (i) infrared thermal sensors that are highly sensitive, (ii) small, lightweight accelerometers and GPS for transportation, motion and position sensing, (iii) sensors to detect biochemicals, (iv) sensors for monitoring health, and (v) sensors related to drug/nutrition delivery systems, (vi) sensors for communication, (vii) sensors for use in integrated circuits and building facilities, (viii) sensors for safety and security, and (ix) sensors for aircraft.

Composition-wise, there can be

(i) Chemical sensors that inco rporate capacitive read-out cantilevers and electronics for signal analysis. They are sensitive enough to detect single chemical and biological molecules. Nano-mechanical cantilever sensors offer low cost, fast response, and high specificity without the need for pre-analysis labeling.
(ii) Electrometers, which consist of a torsional mechanical resonator, a detection electrode and a gate electrode used to couple charge to the mechanical element.
(iii) Optical capillary sensors can be an affordable tool for classifying liquid samples. Moreover, the unique optical properties of noble metal nanoparticles can be used as colorimetric probes for biological analysis. Some important types of nanosensors to aid defense programs are mentioned below.

6.4.1.1 *Blue Crab Nanosensors*

Chitosan is a biodegradable polymer (of chitin monomers) found in the shells of crabs and which is a key component of nanosensor, a "system on a chip." The blue crab nanosensor was developed at the University of Maryland for the detection of minute quantities of explosive, bioagents, chemicals and other dangerous materials in air and water. Chitosan readily binds to negatively charged surfaces.

6.4.1.2 *Nanowire Biosensors*

Nanowire biosensors can detect chemical and biologic material. Biologic materials could be any specific type of DNA in an

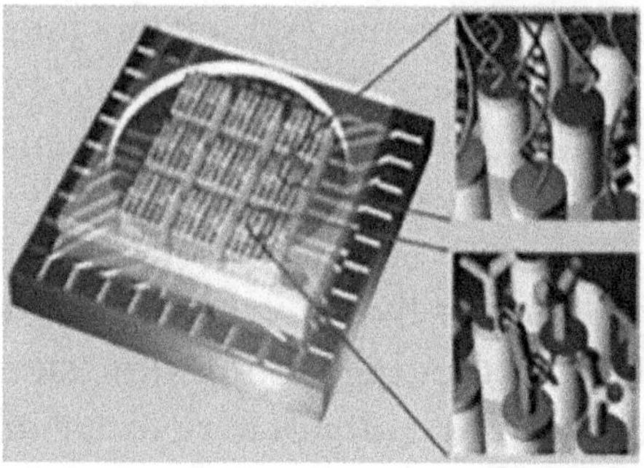

Figure 6.15 DNA molecules attached to the ends of vertical CNT grown on silicon chip.

analyte, any encoded antibody on nano-barcodes, therapeutic serum, toxin, virus, vaccine, blood, blood component or derivatives, allergic products or analogous products or derivatives applicable to the prevention or cure of injuries or disease of humans. There are CNT-based sensors for use with a range of analytes, including gaseous molecules, organic charge transfer complexes, proteins, DNA, and antibodies (Figure 6.15). CNT-based fluidic sensors for studying the shear stress of blood vessels and cells useful in diagnosing many diseases are also being developed.

6.4.1.3 *Intracellular Biosensors*

These biosensing devices could be used to sense pre-malignant cancerous changes in living cells. Such devices are created from dendrimers of nano-diameter.

6.4.1.4 *Biosensors*

Sensors prepared using layer-by-layer self-assembly and the Langmuir–Blodgett technique are highly efficient approaches when working with expensive biological compounds. Because of their unique optical properties, noble metal nanoparticles can be used as colorimetric probes for biological analysis and nanosensors in bioinformatics, which are much needed systems for healthcare.

6.4.1.5 Nanosensor as Nano-Nose

Rather than the response of a single chemical sensor, Prof. Suslick of the University of Illinois has developed a handheld sensor that can be used to detect 19 different toxic chemicals. The array is comprised of different nanoporous pigments with colors that are changed by chemical reaction with the chemical smelled or detected.

6.4.2 Nanotechnology and Protective Clothing for Defense Personnel

Defense personnel are vulnerable to threats from not only advanced lethal weapons, but also from chemical and biological warfare weapons. Hence, they require personal protective equipment such as overalls, boots, hearing protection, face shields, gloves, goggles, insect bars, hats/hamlets, respirator fittings and sensor-attached clothing.

The present-day protective clothing system, such as fabrics made of Kevlar, Nomex and nylon coated with special chemicals to enhance fire-retardant and protective capability, do not meet the requirement to face such biological and chemical weapons. Due to the extraordinary physical, chemical, mechanical, and electrical properties of nanofibers and their composites, nanotechnology offers fabrics for lightweight, flexible, anti-ballistic, comfortable and protective clothing, which incorporates sensors for energy storage, conductive, and self-decontaminating (from biological and chemicals) properties. The use of nanotechnology adaptive nanofibers is being researched that are switchable for thermal control and for camouflaging. Another necessity for the engineered fabrics are that they should be hydrophobic, heat resistant and can be easily molded into a comfortably designed uniform.

6.4.2.1 *Nanofabrics and On-Time Detection and Treatment*

Nanofibers that are a barrier to chemicals are very strong, especially if CNT-Polymer nanocomposites are used, because they have selective insulation and gas permeability.

Nanofibers are synthesized by electrospinning of polymers, such as poly(lactic-co-glycolic) acid (PLGA) and poly(ethylene oxide)-polypyrrole, etc., using high voltage (up to 85 kV) between polymer solution (as source of nanofibers) and a grounded collector plate. The details of nanofabric as smart-skin material are discussed in Chapter 7. Apart from being used as clothing for defense personnel, nanofibers have other uses too, such as (i) for filtration, (ii) in sensor as base because of the electrical resistance of polymer conductive nanofibers, (iii) for insulation, (iv) as reinforcement fiber, (v) for anti-ballistic application, (vi) for selective gas permeation for breathing and protecting from chemical and biological attacks.

Nanofibers of smaller diameters are used for fabricating non-woven mats because the smaller the diameter the greater the tensile strength will be of the nonwoven mats. The increase in tensile strength is due to the increase in number of fiber-to-fiber bonds. Moreover, the polymeric molecules get oriented in the fiber-length direction. A self-assembled honeycomb polyurethane nanofiber has been produced that can trap toxic material [97].

CNT-based fibers have been extensively researched to get fibers of desired mechanical and electrical properties. This is usually done by injecting SWCNTs dispersed in a surfactant solution into a rotating bath of aqueous polymer (PVA) solutions. This mixture of polymer and SWCNTs is then spun into fibers (Figure 6.16). The advantage of using CNTs is that they

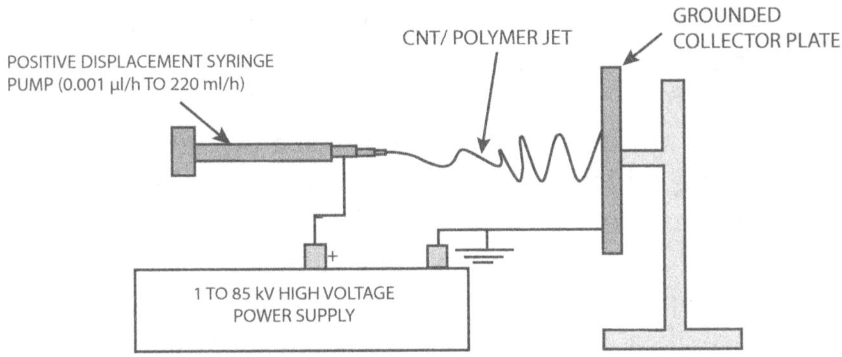

Figure 6.16 Schematic diagram of an electrospinning unit.

can be bent to maximum extent without breaking because their elastic modulus is very high [98].

Moreover, several other nanoparticles like carbon nanofibers, MWCNTs, nano-TiO_2, nano-Al_2O_3, and aluminosilicate nanoclay have also been dispersed in a polymer matrix to produce composites with enhanced mechanical, electrical, and thermal properties.

Utilizing nanotechnology, efforts are underway to develop a smart lightweight suit having properties that can not only protect from bullets, grenade fragments, bioagents and chemical agents; but also offer insulation, ventilation, localized cooling of the soldiers' bodies and are integrated with chemical and biological sensors and nanofiber networks for absorbing and deactivating; along with a helmet for ballistic protection.

To achieve a fabric with the above-mentioned applications, the following developments are taking place:

- ***Camouflage Materials*** are being developed using polymer nanocomposites. One of the achievements in this direction is the development of a high-contrast electromeric nanocomposite material by DeLongchamp *et al.* [99]. This nanocomposite material is based on poly(ethyleneimine) and Prussian blue nanoparticles. It has fully switchable reflective tricolor space coating.

- ***Ballistic Protection Body Armor*** is being developed not only for use as body armor, but also for lining vehicles. CNT is being thought of as a suitable nanomaterial for this. Using nanotechnology, Lee *et al.* [100] have developed a liquid body armor by impregnating Kevlar with colloidal hard spherical silica 450-nm-size nanoparticles dispersed in polyethylene glycol solution, as a shear-thickening fluid (STF). This conjugate greatly enhanced the ability of the material to absorb energy. In the mechanism proposed by Lee *et al.*, at high volume fractions, high shear stresses cause the suspended colloidal particles to form hydro-clusters; leading to an increase in viscosity that can result in a solid-like response (i.e., discontinuous shear thickening). Another nano-conjugate reported by Rosen *et al.* [101] is kaolin clay platelets (about 500 nm diameter) based STF, which may lead to more flexible ballistic armor with reduced total weight. Silicon monoxide (SiO) nanoparticles in a liquid polymer which hardens on ballistic impact (shear thickening fluid), and iron nanoparticles in inert oil which hardens on stimulation with an electrical pulse (magnetorheological fluid), are other possible nanoparticles for use in anti-ballistic body armor.
- ***Fire-Retardant Fabrics*** are being developed in which clay (montmorillonite) nanoparticles are dispersed in polymer matrix to enhance the stiffness, toughness, tensile strength, thermal stability, gas-barrier and flame-retardant properties. These nanocomposites delay the melting, dripping and the rate of burning. Moreover, the clay

nanoparticles improve the mechanical properties; hence, the thickness and weight of the fabric can be significantly reduced. These nanocomposites are either intercalated, in which the polymer or monomers infiltrate the ordered array of silicate layers, or exfoliated, where the silicate layers are delaminated and finely dispersed in the polymer network using dispersion and a compatibilizing agent having one hydrophilic and one organophilic function [102].

- ***Nanocomposite Barrier Fabrics*** are being developed in order to protect soldiers from hazardous biological and chemical warfare agents. Different types of nanocomposites with increased barrier properties have been developed using clay, MgO and TiO_2 in order to protect soldiers and give them an operational advantage [103, 104].
- ***Self-Cleaning, Healing, and Decontaminating Fabrics*** is another priority area for development. TiO_2 nanoparticles are known to have photocatalytic ability to oxidize dirt and other contaminants. To coat TiO_2 on the fabrics (wool, cotton, polyester or polyamide); they are first pre-treated with RF-plasma, MW-plasma or vacuum-UV irradiation to introduce negatively charged carboxylic acid to attach TiO_2. Such TiO_2 conjugated fabrics have self-cleaning ability under sunlight environment [105, 106]. Moreover, Si nanoparticles embedded in epoxy matrix are almost ready to be used within a few years.
- ***Microbial Protective Clothes*** that are self-wound healing, water-repellent, and dirt-repellent have been obtained by coating military garments with silver nanoparticles. Both ionic [107] and silver nanoparticles [108] are well-known antimicrobial materials, especially against infections from

Gram-positive bacteria like *Staphylococcus aureus, Bacillus subtilis,* and Gram-negative *Pseudomonas aeruginosa* and *Escherichia coli.* TiO_2 nanoparticles have also been used as antimicrobial agents in coatings for fabrics [109]. Moreover, antimicrobial drug loading on the textiles is also being tried. It is important that there should be controlled drug release so as to achieve therapeutic efficacy. Use of chitosan for loading and delivering antimicrobial drug has been found to be suitable, as it can release drug from the surface upon contact with the release medium or due to surface erosion. Another way chitosan can release drug is by coming in contact with water, which causes it to swell and release the drug. A ZnO incorporated sodium alginate and PVA composite nanofiber mat has been successfully tried for wound dressing [110].

- ***Chemical and Biological Protective Clothing*** with self-decontamination and drug-delivery properties are being researched. Oxides of nano-metals, such as MgO and Al_2O_3, are good decontaminants. Coating of these nano-metal oxides along with activated carbon and antibiotics are being explored for this purpose [111].
- ***Sensor and Energy-storage Fabrics*** are fabricated for quick on-time action to protect soldiers. Since, devices like sensor, actuator, communication and sighting systems require electrical energy, they are provided with batteries, which add extra weight to the soldiers. The present-day concept is to incorporate piezoelectric devices having nanocrystalline piezoceramic particles into the fabrics. A piezoelectric device can convert mechanical force to electrical signal and vice versa. If incorporated in

a soldier's uniform, it can convert exerted mechanical force from the body, e.g., heart rhythm or pulse, into electrical signals that can be monitored. Some of the promising piezoelectric materials are (i) Poly(vinylidene fluoride) (PVDF) conjugated with clay, CNT, polar matrix (poly acetonitrile), nonpolar matrix (polysulfone) [112, 113]. Attempts are being made to obtain solid piezoelectric film by dip-coating PVDF with acetylene black and $BaTiO_3$ and to obtain porous poly(vinylidene fluoride-trifluoroethylene) film; (ii) Lead zirconate titanate [114, 115]; (iii) Vanadium-doped ZnO nanofiber; (iv) Bi3.15Nd0.85Ti3O12 nanofiber [116]; (v) Barium titanate nanoparticle conjugated with poly(2-hydroxyethl methacrylate) [117].

Using the optical properties of some of the metal oxide semiconductor nanoparticles, such as TiO_2, ZnO nanorods deposited on the flat surface have been used as a biosensor for the detection of dangerous biological compounds such as TiO_2 in bovine leucosis and ZnO nanorods in *Salmonella* [118]. The changes in photoluminescence signal from the nanoparticle surface were applied as biosensor response to detect the analytes. Biosensor based on TiO_2 had a detection range of 2–10 μ g/ml for bovine leucosis antibodies; and based on ZnO had a detection range of 10^2–10^6 cells/ml for *Salmonella* antigen.

A combination of smart fabric and sensor has been developed by Prof. Wang of the University of California, San Diego. He has developed a technique to screen-print sensors on the band of the undergarments of the soldiers, which will remain in close contact to the skin, so that it can monitor the biomarkers in the sweat of the soldier.

6.4.3 Nanorobotics and Other Futuristic Nano-Applications

There are some futuristic ideas that are being thought of but are still in their infancy stage of application for identification and protection to aid biological and chemical warfare defense—nanorobotics is one of them. Moreover, there are many imaginative concepts and fantasies that may become reality in the very near future, some of which are: (i) Use of bacteriophage-based products and techniques for identification of biological pathogens [119]; (ii) Atomic force microscopy as a tool to characterize the physicochemical and mechanical properties of pathogens [120]; (iii) Cerium dioxide (CeO_{2-x}) and orthovanadate ($Gd_{0.9}Eu_{0.1}VO_4$) nanoparticles for protecting a living body from X-ray induced damage [121]; (iv) Various hetero-carbon (heteroatomic derivatives of carbon, in which one or several atoms of carbon are replaced on atoms of another non-metal) nanostructures, such as carbon nitride (g-C_3N_4), azafullerenes, azananotubes (N-dope nanotubes) and N-doped graphene, to be used as very sensitive nanosensors in security systems [122]; (v) Use of spin relaxation in GaAs-based quantum dots for security and quantum information [123], because it is estimated that the relaxation rate in QDs is caused by piezo-phonons.

6.5 Disadvantages of Nanotechnology

Though reducing particle size is an efficient and reliable tool for many applications and nanotechnology has helped in overcoming the limitations of size and has changed the outlook regarding the science; the potential harmful effect of nanomaterial on workers manufacturing nanoparticles should not be overlooked. as they can be inhaled and become lodged in the lungs or get distributed throughout the body. Kwon *et al.* [124] have reported that the inhaled ferro-magnetic NPs can penetrate the BBB and BTB. Some of the medical applications of nanotechnology developed for soldiers would need careful regulation for general medical applications.

6.6 Summary

This chapter has looked into the monstrous biological and chemical warfare that some stubborn leaders and terrorists are bent upon. The possible biological and chemical agents that have been used and that are potential materials to be used in the future are discussed. A brief introduction to the biological materials that include bacteria, fungi, insects, viruses and biogenic toxins are mentioned. Moreover, hazardous and fatal chemical agents have also been touched upon. The chapter finally presents solutions for protecting humans, animals, birds and plants using nanotechnology, which mainly encompasses nanosensors, nanofabrics and futuristic approaches using nanorobotics and other nanomaterials. There is no denying that there are many benefits of nanotechnology in defense, but there are several issues that need careful handling at every stage of development as technology progresses, which is the concluding concern of this chapter.

References

1. Trevisanato, S.I., *Med. Hypotheses*, 69, 6, 1371–1374, 2007, doi: 10.1016/j.mehy.2007.03.012.
2. Mayor, A., *Greek Fire, Poison Arrows, and Scorpion Bombs: Biological and Chemical Warfare in the Ancient World*, pp. 100–101, The Overlook Press, New York, ISBN: 1-58567-348-X, 2003.
3. Rothschild, J.H., *Tomorrow's Weapons: Chemical and Biological*, McGraw-Hill, New York, New York, 1964.
4. Wheelis, M., *Emerg. Infect. Dis.*, 8, 9, 971–975, 2002, doi: 10.3201/eid0809.010536.
5. Calloway, C.G., *The Scratch of a Pen: 1763 and the Transformation of North America (Pivotal Moments in American History)*, p. 73, Oxford University Press, New York, 2007.

6. Warren, C., *J. Aust. Stud.*, 38, 1, 68–86, 2014, doi: 10.1080/14443058.2013.849750#preview.
7. Baxter, R.R., *Buergenthal T. Legal Aspects of the Geneva Protocol of 1925*, Cambridge University Press, Cambridge, England, 2017, doi: 10.2307/2198921.
8. Gold, H., *Unit 731 Testimony (1st ed.)*. Tuttle Pub., New York, 2011.
9. Barenblatt, D., *A Plague upon Humanity: The Secret Genocide of Axis Japan's Germ Warfare Operation*, pp. 220–221, HarperCollins, New York, 2004.
10. Lockwood, J.A., *Six-Legged Soldiers: Using Insects as Weapons of War*, pp. 9–26, Oxford University Press, USA, ISBN: 0195333055, 2008.
11. Institute of Medicine (US) Committee to Review the Health Effects in Vietnam Veterans of Exposure to Herbicides, *Veterans and Agent Orange: Health Effects of Herbicides Used in Vietnam*, vol. 3, National Academies Press (US), Washington (DC), 1994, The U.S. Military and the Herbicide Program in Vietnam. https://www.ncbi.nlm.nih.gov/books/NBK236347.
12. Schwartz, A., War Crimes, in: *The Law of Armed Conflict and the Use of Force: The Max Planck Encyclopedia of Public International Law*, F. Lachenmann and R. Wolfrum (Eds.), p. 1317, Oxford University Press, Oxford, U.K., 2017.
13. Liautard, J.P., Gross, A., Dornand, J., Köhler, S., *Microbiologia*, 12, 2, 197–206, 1996.
14. Chaowagul, W., Simpson, A.J., Suputtamongkol, Y. *et al.*, *Clin. Infect. Dis.*, 29, 2, 375–380, 1999, doi: 10.1086/520218.
15. Dance, D.A.B., Melioidosis and glanders as possible biological weapons, in: *Bioterrorism and Infectious Agents: A New Dilemma for the 21st Century*, I.W. Fong and K. Alibek (Eds.), pp. 99–145, Springer, Boston, MA, 2010, doi: ISBN 978-1-4419-1266-4.
16. Lederberg, J., *Encyclopedia of Microbiology Second Edition A-C*, Volume 1, pp. 781–787, Academic Press, San Diego, CA, 2000.
17. Schachter, J., *N. Engl. J. Med.*, 298, 428–434, 490–496, 540–548, 1978.
18. McPhee, S.J., Erb, B., Harrington, W., *West J. Med.*, 146, 1, 91–95, 1987.

19. MacFarlane, J.T. and Macrae, A.D., *Med. Bull.*, 39, 163–167, 1983.
20. Madariaga, M.G., K, G.M., AWeinstein, R., *Lancet Infect. Dis.*, 3, 1, 709- 721, 2003.
21. Warner, J., Ramsbotham, J., Tunia, E., Valdes, J.J., *Analysis of the Threat of Genetically Modified Organisms for Biological Warfare. Analysis of the Threat of Genetically Modified Organisms for Biological Warfare*, by Center for Technology and National Security Policy National Defense University, Fort Belvoir, VA, 2011.
22. Seshadri, R., Paulsen, I.T., Eisen, J.A., *Proc. Natl. Acad. Sci. U.S.A.*, 100, 9, 5455–5460, 2008, doi: 10.1073/pnas.0931379100.
23. Burrows, W. and Freeman, B.A., *Textbook of Microbiology*, 21st ed., WB Saunders Co, Philadelphia, PA, 1979.
24. Jensen, J.G., Vanderveer, D.E., Greer, W.T., *Analysis for the Automatic In-line Detection of Chemical and Biological Agents in Water Systems*, Chemical Biological Defense Division, Armstrong Laboratory, Int Rpt AL/CF-TR-1996-0181. Wright-Patterson AFB, OH, 1996.
25. Parker, A.C., Kirsi, J., Rose, W.H., Parker, D.T., *Counter Proliferation—Biological Decontamination*, U.S. Army Test and Evaluation Command, Rpt DAAD09-92-D-0004. Aberdeen Proving Ground, MD, 1996.
26. Eitzen, E., Paviin, J., Cieslak, T., Christopher, G., Culpepper, R., *Medical Management of Biological Casualties Handbook*, 3rd ed., U.S. Army Medical Research Institute of Infectious Diseases, Ft Detrick, MD, 1998.
27. White, G.C., *Handbook of Chlorination and Alternative Disinfectants.*, 3rd ed., Van Nostrand Reinhold, New York, 1992.
28. Christopher, G.W., Cieslak, T.J., Paviin, J.A., Eitzen, E.M., Jr., *JAMA*, 278, 5, 412–417, 1997.
29. McGeorge, H.J. II, *Chemical/Biological Terrorism Threat Handbook*, Chemical Research, Development and Engineering Center, Rpt DAAL03-86-D-0001. Aberdeen Proving Ground, MD, 1989.
30. Howard-Jones, N., *Perspect. Biol. Med.*, 16, 292–307, 1973, doi: 10.1353/pbm.1973.0034.
31. Nakase, Y., *Nihon Sakingaku Zasshi*, 50, 637–650, 1995.
32. Cooper, R.C., Olivieri, A.W., Danielson, R.E., Badger, P.G., Spear, R.C., Selvin, S., *Evaluation of Military Field-Water*

Quality. Vol 5: Infectious Organisms of Military Concern Associated with Consumption: Assessment of Health Risks, and Recommendations for Establishing Related Standards, U.S. Army Medical Research and Development Command, APO 82PP2817. Ft. Detrick, MD, 1986.

33. Alchon, S.A., *A Pest in the Land: New World Epidemics in a Global Perspective*, p. 21, University of New Mexico Press, Albuquerque, 2003.
34. Harbeck, M., Seifert, L., Hänsch, S., Wagner, D.M., Birdsell, D., Parise, K.L., Wiechmann, I., Grupe, G., Thomas, A., Keim, P., Zöller, L., Bramanti, B., Riehm, J.M., Scholz, H.C., *PLoS Pathog.*, 9, 5, e1003349, 2013, doi: 10.1371/journal.ppat.1003349
35. Wade, N., *Europe's Plagues Came From China, Study Finds*, New York Times, 31st October, 2010, www.nytimes.com/2010/11/01/health/01plague.html.
36. Grygorczuk, S. and Hermanowska-Szpakowicz, T., *Medycyna Pracy*, 53, 4, 343–348, 2002.
37. McGovern, T.W. and Friedlander, A.M., Plague, in: *Medical Aspects of Chemical and Biological Warfare*, F.R. Sidell, E.T. Takafugi, D.R. Franz (Eds.), pp. 479–502, TMM Publications, Washington, DC, 1997.
38. Klassen-Fischer, M.K., *Clin. Lab. Med.*, 26, 2, 387–95, 2006.
39. Koonin, E.V. and Starokadomskyy, P., *Stud. Hist. Philos. Biol. Biomed. Sci.*, 59, 125–34, 2016, doi: 10.1016/j.shpsc.2016.02.016.
40. Holmes, E.C., *Proc. Natl. Acad. Sci.*, 107, suppl 1, 1742–1746, 2010, DOI: 10.1073/pnas.0906193106Holmes 2007
41. Wimmer, E., Mueller, S., Tumpey, T.M., Taubenberger, J.K., *Nat. Biotechnol.*, 27, 12, 1163–2, 2009, doi: 10.1038/nbt.1593.
42. Asaria, P. and MacMahon, E., *BMJ*, 333, 7574, 890–895, 2006.
43. Bouloy, M. and Flick, R., *Antiviral Res.*, 84, 2, 101–118, 2009.
44. Pepin, M., Bouloy, M., Bird, B.H., Kemp, A., Paweska, J., *Vet. Res.*, 41, 6, 61, 2010.
45. Wilson, M.L., *Ann. N. Y. Acad. Sci.*, 70, 1, 169–179, 2006. http://onlinelibrary.wiley.com.ezp2.lib.umn.edu/doi/10.1111/j.1749-6632, 1994.tb19867.x
46. Feldmann, H., Jones, S., Klenk, H.D., Schnittler, H.J., *Nat. Rev. Immunol.*, 3, 677–685, 2003.

47. Quammen, D., *Insect-Eating Bat May Be Origin of Ebola Outbreak, New Study Suggests*, news.nationalgeographic.com. National Geographic Society, Washington, DC, 2014-12-30, Retrieved 2014-12-30.
48. Angier, N., *Killers in a Cell but on the Loose Ebola and the Vast Viral Universe*, New York Times, 27 October, 2014.
49. Petersen, L.R. and Roehrig, J.T., *Emerg. Infect. Dis.*, 7, 611–14, 2001.
50. Groseth, A., Jones, S., Artsob, H., Feldmann, H., Hemorrhagic fever viruses as biological weapons, in: *Bioterrorism and Infectious Agents: A New Dilemma for the 21st Century*, I.W. Fong and K. Alibek (Eds.), Emerging Infectious Diseases of the 21st Century series, Springer, Boston, MA, 2005.
51. Kiley, M.P., Bowen, E.T., Eddy, G.A., Isaäcson, M., Johnson, K., McCormick, J.B., Murphy, F.A., Pattyn, S.R., Peters, D., Prozesky, O.W., Regnery, R.L., Simpson, D.I., Slenczka, W., Sureau, P., Van Der Groen, G., Webb, P.A., Wulff, H., *Intervirology*, 18, 1–2, 24–32, 1982, doi: 10.1159/000149300.
52. Alibek, K.W. and Handelman, S., *Biohazard: The Chilling True Story of the Largest Covert Biological Weapons Program in the World—Told from the Inside by the Man Who Ran It*, Random House, New York, Delta; Reprint edition (2000).
53. McClain, D.J., Smallpox, in: *Medical Aspects of Chemical and Biological Warfare*, F.R. Sidell, E.T. Takafugi, D.R. Franz (Eds.), pp. 539–559, TMM Publications, Washington, DC, 1997.
54. Dickerson, J.L., *Yellow Fever: A Deadly Disease Poised to Kill Again*, Prometheus Books, Amherst, New York, USA, ISBN:1-59102-399-8,2006,DOIhttps://doi.org/10.1007/0-387-23685-6_6.
55. Croddy, E., Perez-Armendariz, C., Hart, J., *Chemical and Biological Warfare: A Comprehensive Survey for the Concerned Citizen*, pp. 30–31, Springer-Verlag, Heidelberg, ISBN: 0387950761, 2002.
56. Alyokhin, A., Baker, M., Mota-Sanchez, D., Dively, G., Grafius, E., *Am. J. Potato Res.*, 85, 395–413, 2008.
57. Farhang-Azad, A., Traub, R., Baqar, S., *Science*, 227, 543–545, 1985.

58. Ostrolenk, M. and Welch, H., *Am. J. Public Health Nations Health*, 32, 487–494, 1942.
59. Levine, O.S. and Levine, M.M., *Rev. Infect. Dis.*, 13, 4, 688–696, 1991.
60. Förster, M., Klimpel, S., Sievert, K., *Vet. Parasitol.*, 160, 163–167, 2009.
61. Georghiou, G.P. and Hawley, M.K., *Bull. World Health Organ*, 45, 43–51, 1971.
62. Keiding, J., Problems of housefly (Musca domestica) control due to multiresistance to insecticides. *J. Hyg. Epidemiol. Microbiol. Immunol.*, 19, 340–355, 1975.
63. Williams, R.E., *Veterinary Entomology, Livestock and Companion Animals*, CRC Press, Boca Raton, FL, USA, 2010.
64. Euzéby, J.P., *Int. J. Syst. Bacteriol.*, 47, 590–592, 1997.
65. Unsworth, N.B., Stenos, J., Graves, S.R., Faa, A.G., Cox, G.E., *Emerg. Infect. Dis.*, 13, 566–573, 2007.
66. World Health Organization and the Special Programme for Research and Training in Tropical Diseases (TDR), *Dengue Guidelines for Diagnosis, Treatment, Prevention and Control*, World Health Organization, Geneva, 2009.
67. Worrall, E., Basu, S., Hanson, K., *Trop. Med. Int. Health*, 10, 1047–1059, 2005.
68. William, R.H., *An Evaluation of Entomological Warfare as a Potential Danger to the United States and European NATO Nations*, U.S. Army Test and Evaluation Command, Dugway Proving Ground, via thesmokinggun.com, 1981.
69. Greene, E., *Science*, 243, 643–646, 1989.
70. Redd, J.T., Voorhees, R.E., Török, T.J., *J. Am. Acad. Dermatol.*, 56, 952–955, 2007.
71. Irwin, M.E. and Kampmeier, G.E., Commercial products, from insects, in: *Encyclopedia of Insects*, V.H. Resh and R. Carde (Eds.), p. 6, Academic Press, San Diego, via University of Illinois and Illinois Natural History Survey, 2008.
72. Service, M.W., *Medical Entomology for Students*, pp. 81–92, Cambridge University Press, Cambridge, United Kingdom, 2008.
73. Carpenter, S., Groschup, M.H., Garros, C., Felippe-Bauer, M.L., Purse, B.V., *Antiviral Res.*, 100, 102–113, 2013.

74. Linley, J.R., *J. Med. Entomol.*, 22, 589–599, 1985.
75. Kazimírová, M., Sulanová, M., Kozánek, M., Takác, P., Labuda, M., *Haemostasis*, 31, 294–305, 2001.
76. Wilkerson, R.C., Butler, J.F., Pechuman, L.L., *Myia*, 3, 515–546, 1985.
77. Piper, R., *Extraordinary Animals: An Encyclopedia of Curious and Unusual Animals*, Greenwood Press, Westport, Conn., 2007.
78. Kalelioğlu, M., Aktürk, G., Aktürk, F., Komsuoğlu, S.S., Kuzeyli, K., Case report. *J. Neurosurg.*, 71, 929–931, 1989.
79. Lagacé-Wiens, P.R.S., Dookeran, R., Skinner, S., Leicht, R., Colwell, D.D., *Emerg. Infect. Dis.*, 14, 64–66, 2008.
80. Curran, J., *"Screw-Worm Fly"*, Department of Agriculture Farmnotes, Government of Western Australia, 2002.
81. James, M.T., *The Flies that Cause Myiasis in Man*, USDA Miscellaneous Publication No. 631, U.S. Dept. of Agriculture, Washington, D.C., 1947.
82. Franz, D.R., Defense against toxin weapons, in: *Medical Aspects of Chemical and Biological Warfare*, F.R. Sidell, E.T. Takafugi, D.R. Franz (Eds.), pp. 603–619, TMM Publications, Washington, DC, 1997.
83. Vuori, E., Himberg, K., Waris, M., Niinivaara, K., Drinking water purifiers in removal of hepato- and neurotoxins produced by cyanobacteria, in: *Recent Advances in Toxinology Research.*, vol. 3, P. Gopalakrishnakone and C.K. Tan (Eds.), pp. 318–322, National University of Singapore, Venom and Toxin Research Group, Singapore, 1992.
84. Warner, J.S., *Review of Reactions of Biotoxins in Water. Rpt CBIAC Task 152*, U.S. Army Medical Research and Development Command, Ft. Detrick, MD, 1990.
85. Marshall, E., *Science*, 275, 745, 1997.
86. Zilinskas, R.A., *JAM*, 278, 5, 418–424, 1997.
87. Ulrich, R.G., Sidell, S., Taylor, T.J., Wilhelmsen, C.L., Franz, D.R., Staphylococcal enterotoxin B and related pyrogenic toxins, in: *Medical Aspects of Chemical and Biological Warfare*, F.R. Sidell, E.T. Takafugi, D.R. Franz (Eds.), pp. 621–630, TMM Publications, Washington, DC, 1997.

88. Palmgren, M.S. and Ciegler, A., Aflatoxins, in: *Handbook of Natural Toxins. Vol 1: Plant and Fungal Toxins*, R.F. Keeler and A.T. Tu (Eds.), pp. 299–323, Marcel Dekker, New York, 1983.
89. Simon, J.D., *JAMA*, 278, 5, 428–430, 1997.
90. Gupta, R., *Handbook of Toxicology of Chemical Warfare Agents*, Academic Press, Boston, 2009.
91. Whalley, C.E., *Toxins of Biological Origin. Rpt CRDEC-SP-021, AD B145632*, Chemical Research, Development and Engineering Center, Aberdeen Proving Ground, MD, 1990.
92. Wannemacher, R.W. Jr, Dinterman, R.E., Thompson, W.L., Schmidt, M.O., Burrows, W.D., *Treatment for Removal of Biotoxins from Drinking Water. Rpt no TR9120, AD A275958*, US Army Biomedical Research and Development Laboratory, Ft. Detrick, MD, 1993.
93. Stewart, C.E., *Weapons of Mass Casualties and Terrorism Response Handbook*, p. 175, Jones & Bartlett Learning, Sudbury, 2006.
94. Wheelis, M., Rózsa, L., Dando, M. (Eds), Deadly Cultures: Biological Weapons Since 1945, Harvard University Press, Cambridge, MA, USA, 2006.
95. Woodward, R.B., *Pure Appl. Chem.*, 9, 1, 49–75, 1964, doi: 10.1351/pac196409010049.
96. Auerbach, P.S., Clinical therapy of marine envenomation and poisoning, in: *Handbook of Natural Toxins. Vol 3: Marine Toxins and Venoms*, A.T. Tu (Ed.), pp. 493–565, Marcel Dekker, New York, 1988.
97. Thandavamoorthy, S., Gobinath, N., Ramkumar, S.S., *J. Appl. Polym. Sci.*, 101, 5, 3121–124, 2006.
98. Vigolo, B., Penicaud, A., Coulon, C., Sauder, C., Rene, P., Journet, C., Bernier, P., Poulin, P., *Science*, 290, 5495, 1331–334, 2000.
99. DeLongchamp, D.M., Kastantin, M., Hammond, P.T., *Chem. Mater.*, 15, 8, 1575–1586, 2003, doi: 10.1021/cm021045x.
100. Lee, Y.S., Wetzel, E.D., Wagner, N.J., *J. Mater. Sci.*, 38, 13, 2825–833, 2003.
101. Rosen, B.A., Nam Laufer, C.H., Kalman, D.P., Wetzel, E.D., Wagner, N.J., Multi-threat performance of Kaolin based shear

thickening fluid (STF)-treated fabrics, in: *52nd International SAMPE Symposium and Exhibition*, Baltimore, 2007.

102. Kashiwagi, T., Grulke, E., Hilding, J., Groth, K., Butler, K.G., Shields, J., Kharchenk, S., Douglas, J., *Polymer*, 45, 4227–239, 2004.
103. Rajagopalan, S., Koper, O., Decker, S., Klabunde, K.J., *Chem.-Eur. J.*, 8, 2602–607, 2002.
104. Watson, S., Beydoun, D., Scott, J., Amal, R., *J. Nanoparticle Res.*, 6, 193–207, 2004.
105. Bozzi, A., Yuranova, T., Kiwi, J., *J. Photochem. Photobiol. A*, 172, 1, 27–34, 2005.
106. Yuranova, T., Mosteo, R., Bandara, J., Laub, D., Kiwi, J., *J. Mol. Catal. A-Chem.*, 244, 1-2, 160–67, 2006.
107. Yang, F.C., Wu, K.H., Huang, J.W., Horng, D.N., Liang, C.F., Hu, M.K., *Mater. Sci. Eng. C.*, 32, 5, 1062–67, 2012.
108. Zhang, L., Luo, J., Menkhaus, T.J., Varadaraju, H., Sun, Y., Fong, H., *J. Membr. Sci.*, 369, 1, 499–05, 2011.
109. Pant, H.R., Pandeya, D.R., Nam, K.T., Baek, W.I., Hong, S.T., Kim, H.Y., *J. Hazard. Mater.*, 189, 1, 465–71, 2011.
110. Shalumon, K.T., Anulekha, K.H., Nair, S.V., Chennazhi, K.P., Jayakumar, R., *Int. J. Biol. Macromol.*, 49, 3, 247–54, 2011.
111. Munim, M. and Ramkumar, S.S., *Indian J. Fibre Text. Res.*, 31, 41–51, 2006.
112. He, L., Xu, Q., Hue, C., Song, R., *Polym. Compos.*, 31, 5, 921–27, 2010.
113. Kim, Y.-H., *J Korean Inst. Electr. Electron. Mater. Eng.*, 23, 3, 250–259, 2010.
114. Dharmaraj, N., Kim, C. H., Kim, H. Y. *Mater. Lett.*, 59, 3085–3089, 2005.
115. Xu, S.Y., Shi, Y., Kim, S.G., *Nanotechnol*, 17, 4497–01, 2006.
116. Liao, M., Zhong, X.L., Wang, J.B., Xie, S.H., Zhou, Y.C., *Appl. Phys. Lett.*, 96, 012904, 2010.
117. Mimura, K.I., Moriya, M., Wataru, S., Toshinobu, Y., *Compos. Sci. Technol.*, 70, 3, 492–497, 2010.
118. Tereshchenko, A.V., Smyntyna, V.A., Konup, I.P., Geveliuk, S.A., Starodub, M.F., Metal oxide based biosensors for the detection of dangerous biological compounds, in: *Nanomaterials for Security*, J. Bonča and S. Kruchinin (Eds.), NATO Science for Peace and

Security Series A: Chemistry and Biology, Springer, Dordrecht, 2016, DOI https://doi.org/10.1007/978-94-017-7593-9_22.

119. Chanishvili, N., Bacteriophage-based products and techniques for identification of biological pathogens, in: *Nanotechnology to Aid Chemical and Biological Defense*, T. Camesano (Ed.), NATO Science for Peace and Security Series A: Chemistry and Biology, Springer, Dordrecht, 2015, DOI https://doi.org/10.1007/978-94-017-7218-1_2.

120. El-Kirat-Chatel, S. and Beaussart, A., Atomic force microscopy tools to characterize the physicochemical and mechanical properties of pathogens, in: *Nanotechnology to Aid Chemical and Biological Defense*, T. Camesano (Ed.), NATO Science for Peace and Security Series A: Chemistry and Biology, Springer, Dordrecht, 2015, DOI https://doi.org/10.1007/978-94-017-7218-1_1.

121. Grygorova, G., Klochkov, V., Mamotyuk, Y., Malyukin, Y., Cerium dioxide CeO2-x and orthovanadate (Gd0.9Eu0.1VO4) nanoparticles for protection of living body from X-ray induced damage, in: *Nanomaterials for Security*, J. Bonča and S. Kruchinin (Eds.), NATO Science for Peace and Security Series A: Chemistry and Biology, Springer, Dordrecht, 2016, DOI https://doi.org/10.1007/978-94-017-7593-9_23.

122. Kharlamova, G., Kharlamov, O., Bondarenko, M., Khyzhun, O., Hetero-carbon nanostructures as the effective sensors in security systems, in: *Nanomaterials for Security*, J. Bonča and S. Kruchinin (Eds.), NATO Science for Peace and Security Series A: Chemistry and Biology, Springer, Dordrecht, 2016, DOI https://doi.org/10.1007/978-94-017-7593-9_19.

123. Prabhakar, S. and Melnik, R., Spin relaxation in GaAs based quantum dots for security and quantum information processing applications, in: *Nanomaterials for Security*, J. Bonča and S. Kruchinin (Eds.), NATO Science for Peace and Security Series A: Chemistry and Biology, Springer, Dordrecht, 2016, DOI https://doi.org/10.1007/978-94-017-7593-9_3.

124. Kwon, S.K., Ahn, J.M., Kim, G.H., Chun Chang, Hwan., *Polym. Eng. Sci.*, 42, 11, 2165–2171, 2002.

7

Smart Nanofabrics for Defense

Madhuri Sharon

Walchand Centre for Research in Nanotechnology & Bionanotechnology, WCAS, W.H. Marg, Ashok-Chowk, Solapur, Maharashtra, India

Any sufficiently advanced technology is indistinguishable from magic.
Arthur C. Clarke

7.1 Introduction

With the advent of nanotechnology, the potential for revolutionizing its applications in many fields was sought. Development of smart nanofabrics or nanotextiles is one of them. It caught the eye of defense researchers also because at nanoscale, materials can be manipulated to result in new functionalities for intelligent textiles that can self-clean, sense, actuate, and even communicate. Materials like antimicrobial nanoparticles for fabric surface coating and inherently conducting polymers like carbon nanotubes, etc., result in many possible applications for the military, healthcare, sports, and new fashion trends. For military applications, it is being envisaged that these fabrics can monitor the hazards of the wearer's surrounding environment and give continuous updates on the health status of an

Madhuri Sharon, et al. Nanotechnology in the Defense Industry: Advances, Innovation, and Practical Applications, (235–274) © 2019 Scrivener Publishing LLC

individual; smart textiles can have body sensor networks incorporating sensors, actuators, control, and wireless data transmission devices, etc. The challenges that smart textiles face are not only improvising and advancing the current functionalities of textiles; but also maintaining the look, tactility, flexibility, comfort, washability and the feel of the fabric. For smart textiles use in the military, it is imperative to integrate thermoregulators to maintain body temperature, sensors to monitor the wearer and the environment; as well as have a device to process the information, sense and react to stimuli (mechanical, thermal, chemical, electrical, or magnetic sources). So, in a nutshell, smart fabrics should have sensors, actuators, control units, data transmission, and processing units integrated into them. To integrate all these functionalities the key component is application and use of nanotechnology.

7.2 A Brief History of Smart Skin Material

Improvisation of basic fabric using conductive threads is a nearly 1,000-year-old effort. For centuries artisans from India, China, Greece, Byzantium, Japan and Korea were wrapping very fine gold and silver thread around fabric to create brocade [1]. About 3,000 years ago, Rishi Atri and his wife Anusuya presented Sita (Ram's wife) with garments (saree) that was self-cleaning [2]. The manufacturing of brocade began in China during the period from 481 BC to 403 BC [3]. Zari brocade (a type of silk brocade with silver, golden and copper threads) was developed in India during emperor Akbar's reign in the mid-sixteenth century.

At the end of the 19th century, clothing combined with electricity was developed to illuminate fabric [4]. By the 1960s, the first astronaut spacesuit, along with clothing that could inflate and deflate, light up, and heat and cool itself, was made for astronaut Alan Shepard when he flew on May 5, 1961; and in 1985

Harry Wainwright [5] created a fully animated sweatshirt for the general public consisting of fiber optics, LED and a microprocessor to control individual frames of animation, resulting in a full color cartoon on the surface of apparel. Later, in 1995, he invented the first machine enabling fiber optics to be machined into fabrics in 1997. It is worth mentioning that German machine designer, Herbert Selbach of Selbach Machinery produced the first CNC machine that could automatically implant fiber optics into any flexible material and produced animated coats for Disney Parks in 1998. The latest smart fabric was produced in 2005 and displayed at the Smart Fabrics Conference held in Washington D.C. on May 7th, 2007. Also demonstrated at the conference was an ECG Bio-Physical display jacket having LED/Optic displays using GSR sensors in a watch connected via Bluetooth to the embedded machine washable display in a denim jacket. Another milestone in smart textiles was Wainwright's presentation of Infra-Red digital displays machine embedded into fabrics for IFF (identification of friend or foe), which won an "Honorable Mention" award from NASA in 2010. This product was the impetus for MIT scientists to work on "Wearable Computers" and the trend of integrating electronic devices into clothing evolved. In 2012, Wainwright demonstrated smart fabric creations that change color using any smart phone, indicate callers on mobile phones without a digital display, and contain WIFI security features that protect purses and personal items from theft. Moreover, the possibility of integrating digital electronics with conductive fabrics and a method for embroidering electronic circuits are also being explored [6].

7.3 Types of Smart Textiles

Based on the improvisation of smart textiles to date, the three generations of smart textiles are classified as (i) Passive smart textiles, (ii) Active smart textiles, and (iii) Ultra-small textiles.

7.3.1 Passive Smart Textiles

Passive smart textiles are the first generation of smart textiles. The new materials added to the textiles enable passive, noninvasive sensing of wearers and their environs [7]. Passive mode works irrespective of changes in the environment [8]. For example, an insulating coat will maintain the temperature to the same degree irrespective of the outside temperature. Other passive smart textile attributes are anti-microbial, anti-odor, anti-static and bulletproof capabilities among others.

7.3.2 Active Smart Textiles

Active smart textiles are second generation smart textile. They contain both actuators and sensors. In the active mode of working these active textiles adapt their functionality to changing environment automatically. Active textiles are water resistant (hydrophilic), permeable to vapor and vapor absorbing, thermo-regulated garments that maintain the wearer's body temperature, can store and evolve heat, the fabric (suit) is electrically heated, they are shape memory and chameleonic.

7.3.3 Ultra Small Textiles

Ultra small textiles are smartest third generation intelligent textiles. Not only can they sense, but they can also react and adapt to changing environmental conditions or stimuli by reasoning and activating responses. Present-day suits of ultra-smart textiles are composed of traditional textiles and clothing technology integrated with modern scientific developments, such as structural mechanics, sensor, actuator, advance processing and communication technology, along with artificial intelligence pertaining to their biological applications. One of the major inputs in these textiles is miniaturized electronic components, which is a new material for suitable fabrics. They are flexible, comfortable to wear

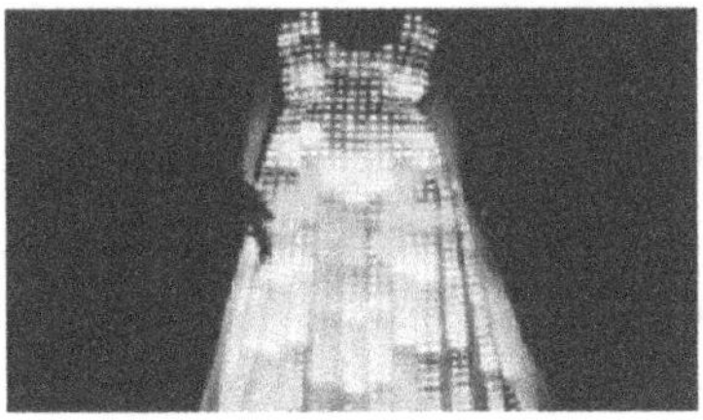

Figure 7.1 Light-emitting smart textile.

and can be designed per requirement, such as light-emitting fabric (Figure 7.1).

7.4 Fabrication of Smart Textiles

In this section we will discuss smart functional materials and fibers used in smart textiles. Cotton, wool and flax are the earliest known materials used for weaving fabrics. Then came the era of synthetic fabrics; and now a fabric known as Kevlar®, trademarked name of poly-*para*-phenylene terephthalamide, is used in bulletproof vests because of its ultrahigh strength. Titanium is strong and very light and a titanium plate can withstand little pellet guns all the way up to huge armor piercing .50 cal black tip bullets.

Another development is smart nanotextile which is made of conducting polymers as the base material interwoven with carbon nanotubes (CNTs) or carbon coated threads of conductive yarn, conductive rubber, and a number of other materials in the form of nanoparticles or nanofibers for different functionalities. The needs of today's military personnel and their safety demand fabrics having mobility, healthcare or rehabilitation, novel sensors and actuators integrated into textiles. The materials of smart textiles that are being fabricated have to be able to interact, communicate and sense.

7.4.1 Metal Fibers

Very thin pure metallic threads are made of composite stainless steel or a fine continuous conductive metal-alloy combination of fibers with conductive materials. The processes used for producing these fibers are (i) through a bundle-drawing or (ii) by shaving off the edge of thin metal sheeting. These fibers are used for making fabric by weaving or knitting; using electrodes for monitoring electrical physiological signals.

7.4.2 Conducting Inks

Prior to development of conductive ink, the following materials were known to be used: (a) ***Carbon ink*** made of charcoal or lamp black mixed with a water soluble gum as adhesive; (b) ***Metal (Iron) Gall Ink*** made of tannic acids and ferrous sulphate (copperas) solution and gum thickeners; (c) ***Black Ink*** made of carbon black; (d) ***White Ink*** using titanium dioxide as pigment; nowadays it contains additives such as waxes, lubricants, surfactants, and drying agents to aid printing; (e) ***Specialized Inks*** use dyes with special characteristics and colors in aqueous or organic solvents (propylene glycol, propyl alcohol, toluene, glyco ethers, etc.) and resins, along with preservatives and wetting agents; (f) ***Silver ink*** has found many applications today such as printing RFID (radio frequency identification) tags in transit tickets; to improvise or repair circuits of printed circuit boards; computer keyboards with printed circuits that sense when a key is pressed; windshield defrosters having resistive traces of ink printed on the glass and, lately, conductive traces are printed on rear windows of cars to function as the radio antenna.

Conductive ink is typically created by infusing graphite or other conductive materials into ink and is used for printing objects that can conduct electricity. In smart textile, this ink is used in sensors and/or as an interconnection substrate.

Conductive inks have replaced traditional industrial standards such as etching copper from copper-plated substrates to form the same conductive traces on required substrates. Conductive ink is being thought of as sound controllers attached to the textile, which could be activated when the user presses it. By adding carbon, copper, silver, nickel and gold to conventional printing inks, they can be made conductive. Also, a desired layout can be screen-printed using conductive inks to add conductivity to specific areas of textile. Printed areas can be subsequently used as switches or pressure pads for the activation of circuits.

7.4.3 Inherently Conductive Polymers (ICP)

Inherently conductive polymers (Figure 7.2) are used in smart textiles because they have sensing as well as actuation properties. The first ICP (doped polyacetylene) was discovered in 1977 by Chiang *et al.* It could conduct electricity and had the ability to sense and actuate. Later, in 1996 Baughmann *et al.* [10] studied actuators based on ICPs that generated much higher stresses than natural skeletal muscle and higher strains than high-modulus ferroelectrics and called it artificial muscle. Huang *et al.* [11] showed that the sensors based on ICPs change their resistivity or generate an electrical signal in response to external stimuli. Conductive yarns and fibers are made by mixing pure metallic or natural fibers with conductive materials. To make a fiber conductive, fibers are filled with conductive material (e.g., carbon or metallic particles) or coated with conductive polymers or metal and fibers spun with thin metallic or plastic conductive threads. Fabrication of ICP is done by chemical or electrochemical oxidation of monomers. Oxidation is done either in solution or in the vapor phase [12]. Usually dopants are incorporated in ICP to enhance the required properties. ICP textile fibers are also made by continuous wet spinning.

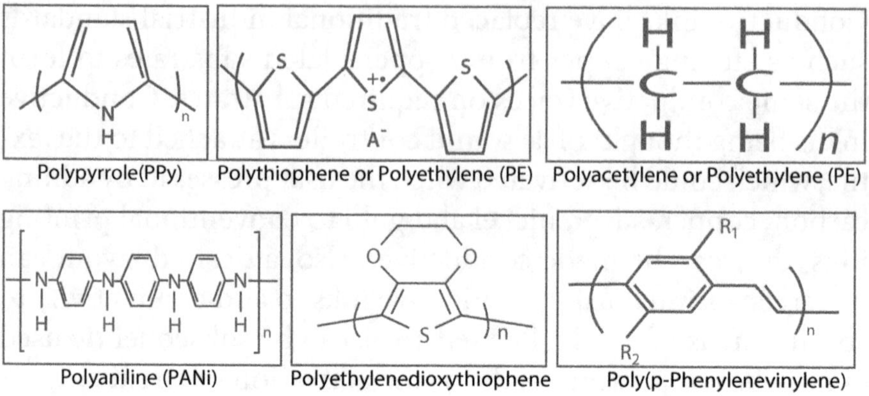

Figure 7.2 Some conductive polymers.

Some of the common ICP-based smart polymers (in undoped form) that can sense, process information and actuate are shown in Figure 7.2 and are briefly described below.

7.4.3.1 Polypyrrole (PPy)

Polypyrrole has high mechanical strength and high elasticity; and is relatively stable in air and electroactive in both organic and aqueous solutions.

7.4.3.2 Polyacetylene or Polyethylene or Polythene (PE)

Since polyacetylene is unstable in air it has limited uses.

7.4.3.3 Polyaniline (PANi)

Polyaniline has relatively good environmental stability and good electrical conductivity. PANi exists in three oxidation states: (i) Fully reduced state, i.e., leucoemeraldine base, (ii) Partly oxidized state, i.e., emeraldine base, and (iii) Fully oxidized state, i.e., pernigraniline base. Another quality of PANi is that when it is oxidized in aqueous protonic acids, PANi increases its conductivity by 9–10 orders of magnitude.

Thin films of PANi/CSA nanofibers with diameters between 30 and 50 nm have been used as a chemical sensor [13] with excellent performance in sensing vapors of acid (HCl) as well as base NH_3, which was possibly because of their diameters. Films of ICP nanofibers are used as sensors to detect chemical vapors that interact with ICPs and change their conductivity. One more advantage of the ICP material is that sensors retain the natural texture of the material. However, prolonged use of this device tends to show a variation in resistance over time and high thus affecting the response time.

7.4.3.4 *Polythiophene and Its Derivatives*

Polythiophene and its derivatives have both *p*- and *n*-type electronic forms, which are used for constructing polymer field-effect transistors for flexible logic circuits [14]. It is also used in solar cell [15, 16].

7.4.4 Electrically Conductive Polymers (ECP)

Electrically conductive polymers possess electronic conductivity variable over many orders of magnitude, ion-transport capability, junction effects and electrode effects allied to polymeric physical properties. Hence, they have a wide range of applications (Figure 7.3) in the electronics industry. Conventionally, electrically conductive textiles are used for controlling static

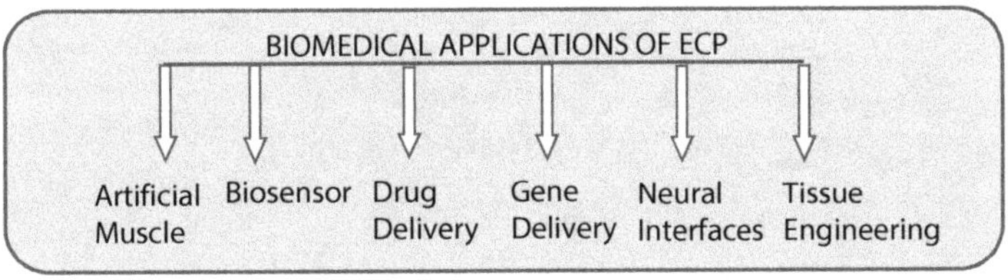

Figure 7.3 Various biomedical applications of electrically conductive polymers.

and electromagnetic interference shielding. But now electrically conductive smart textiles are used as electrodes as well as interconnection between different components.

7.4.5 Optical Fibers

Optical fibers are made of glass (silica) or plastic. They are very thin, flexible, and transparent. An optical fiber has a core surrounded by a transparent cladding material (Figure 7.4) with a lower index of refraction. They are used in smart garments for (i) transmitting data signal, (ii) transmitting light for optical sensing, (iii) detecting deformations in fabrics due to stress and strain, (iv) performing chemical sensing, and (v) data transmission. Plastic optical fibers are easily integrated into a textile; they do not generate heat and are insensitive to electromagnetic radiation. There are optical fibers capable of emitting their own light which are used in safety vests.

When light travels through the fiber-optic cable the internal reflection keeps light rays bouncing down the inside of a fiber-optic cable and repeatedly bouncing off the walls. Apart from smart textiles, optical fibers are used for high-speed

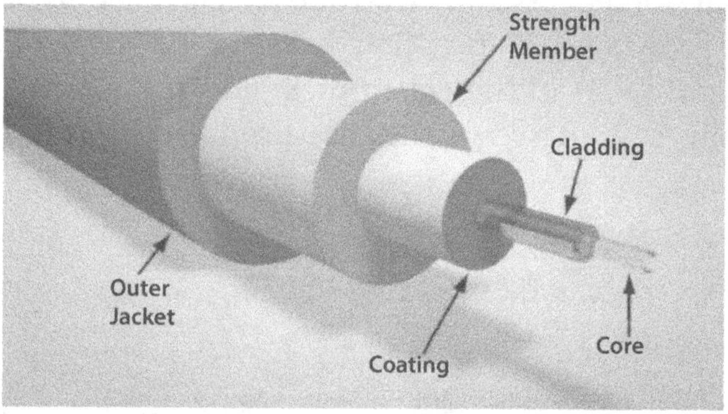

Figure 7.4 Typical fiber-optic cable.

communications such as cable TV and high-speed broadband services. Information in the form of pulses of light is sent down bundles of optical fibers. Fiber-optic cables are able to carry more signals than traditional copper cable telephone lines.

7.4.6 Shape Memory Material (SMM)

Shape-memory alloys are materials that change their shape due to external stimulus; however, they have the ability to remember their original shape and when a particular stimulus, say heat, is applied above a certain temperature, they return to their pre-deformed shape. This is known as the shape memory *effect* (SME). A shape-memory alloy (SMA) is a lightweight, solid-state alternative to conventional actuators such as hydraulic, pneumatic, and motor-based systems. The two most prevalent SMAs are copper-aluminum-nickel and nickel-titanium (NiTi) (Figures 7.5 and 7.6); however, SMAs can also be created by alloying zinc, copper, gold and iron. Nickel titanium, also known as Nitinol, is part of shape memory alloy. Nitinol has many applications such as dental braces, spectacle frames and stents.

The SMAs provide increased protection against sources of heat. Properties of SMA differ at below and above temperature at which it is activated, because the activation temperature depends on the ratio of nickel to titanium in the alloy. Similarly, cuprous-zinc alloys are capable of producing the reversible variation needed for protection from changeable weather conditions. Shape memory polymers (SMPs) have the same effect as the Ni-Ti alloys, but, being polymers, they are more suited to textiles. Electroactive polymers (EAPs) made up of highly functionalized polymer gel robots (poly 2-acrylamido-2-methylpropane sulfonic acid) are expected to replace muscles and tendons.

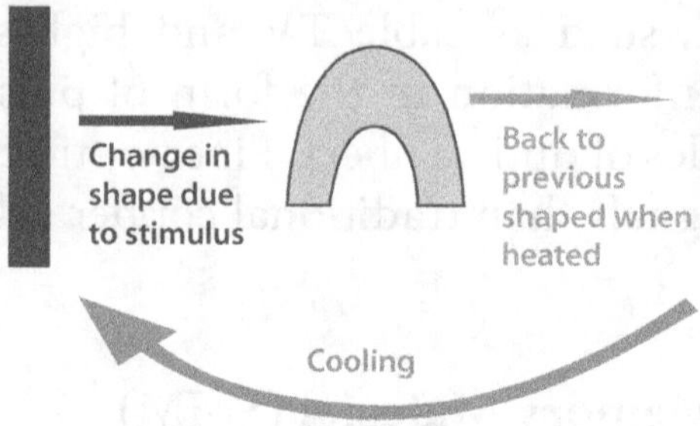

Figure 7.5 Illustration of SMA deformation and strain recovery. (Here SMA is in martensite form at lower temperatures and when heated gets transformed into austenite form.)

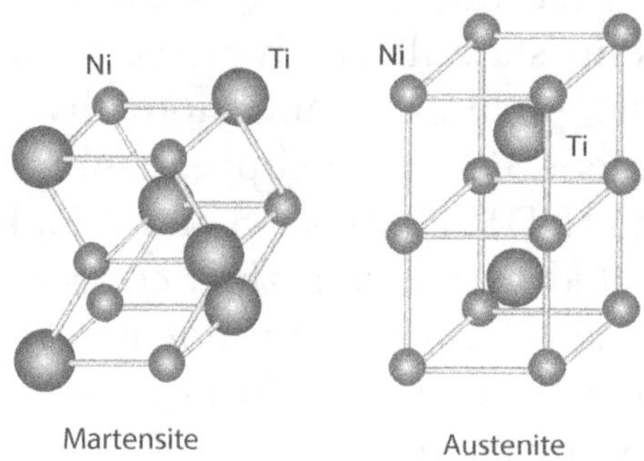

Figure 7.6 Shape memory materials.

7.4.7 Chromic Material

When light interacts with material a color is produced, which is called chromic phenomena. The changes in color of the compound are often reversible. Chromism is based on a change in the π- or d-electron state of molecules. External stimuli are responsible for altering the electron density of the substance. There are many natural as well as artificial compounds with specific chromism. Chromic materials are categorized under the following five headings.

7.4.7.1 Thermochromism

Thermochromism, as its name suggests, is induced by heat. It is the property of substances to change color due to a change of temperature. Here heat is the external stimulus.

7.4.7.2 Photochromism

Photochromism is induced by a change in light intensity, e.g., light-sensitive glasses which darken in the sun. Photochromic inks are used to make photochromic material because they are unstable and exist as powdered crystal. This is due to the isomerization between two different molecular structures, light-induced formation of color centers in crystals and precipitation of metal particles in glass (Figure 7.7).

7.4.7.3 Electrochromism

Electrochromism is induced by the gain and loss of electrons in compounds with redox active sites, such as metal ions or organic radicals, due to electricity as external stimulus. Electrochromism is a phenomenon where a material changes

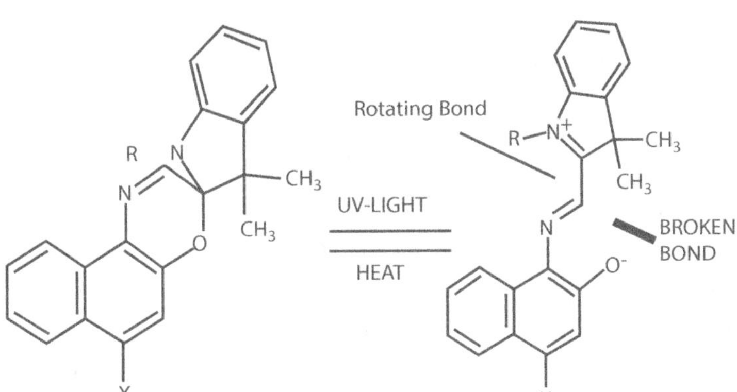

Figure 7.7 Schematic of changes in the chemical structure of a photochromic material in order to change color.

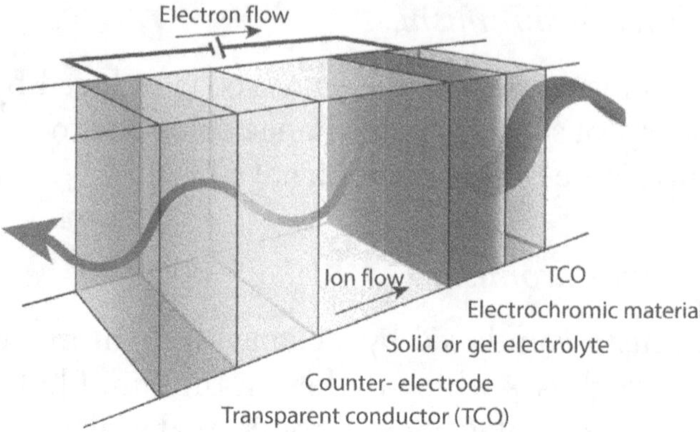

Figure 7.8 Schematic of electrochromic device stack, depicted in a darkened state. Electrons flow through an external circuit into the electrochromic material, while ions flow through the electrolyte to compensate for the electronic charge. (Source: Runnerstrom *et al.* [17]. Open Access Article, This Open Access Article is licensed under a Creative Commons Attribution 3.0 Unported License)

color, transparency, or other optical property in response to electric charge (Figure 7.8). In a nutshell, the electrochromic phenomenon is displayed by some materials of reversibly changing color stimulated by redox reactions. These electrochromic displays are based on any material that changes color depending on the applied potential.

7.4.7.4 Piezochromism

Piezochromic solid or liquid material changes color in response to external pressure stimulus. Plastics and metal complexes, such as Ni(II), Pd(II) and Pt(II) complexes with dimethylglyoximate, L1: dmg, are known to have this quality, which is explained by the pressure perturbation to the HOMO (highest occupied molecular orbital) and/or LUMO (lowest unoccupied molecular orbital) energy levels of the related electronic transition. These properties of the solids, i.e., tuning the external pressure, are being used as electronic devices and as pressure sensors. These metal complexes can be used for *in-situ* pressure sensors or as conductors in extreme environments.

7.4.7.5 Solvation Chromism

The external stimulus for solvate chromic materials is liquid or gas. Solvation chromism is the absorption and emission of a chemical species induced by the solvents and depends on the polarity of the solvent. Most solvate chromic compounds are metal complexes. It is a reversible process that is used for probes for application in polymer characterization.

7.4.8 Phase Change Materials (PCM)

In defense, PCMs are used in textiles for antiballistic vests, automotive applications, medical devices incorporated in apparel such as underwear, socks and shoes, bedding accessories, sleeping bags, etc. It is imperative that the PCM incorporated in textiles be the temperature range of human skin, so that the protective garments will be suitable in all types of weather, whether the coldest winter or hottest summer. Smart PCM incorporated textile materials can store excess heat and release it back when the heat is needed. PCMs can be applied during coating as well as during fiber spinning or chemical finishing like lamination.

7.5 Nanoparticle Coated Textiles

The basic materials used to fabricate smart textiles are metallic silk, organza, stainless steel filament, metal clad *aramid* fiber (synthetic polyamide fibers formed from aromatic monomers of high strength and thermal stability), conductive polymer fiber, conductive polymer coating and special carbon nanomaterials as fiber. The basic fabric material is decided on as per its use. Carbon nanofibers (CNF) are used as filler to increase tensile strength and chemical resistance conductivity. Nanotechnology involving the use of nanoparticles is a growing interdisciplinary technology that is being applied to enhance the desired properties and applications of smart textiles. As per demand and expectations, modern-day smart textiles have to be comfortable, hygienic, stain resistant, colorfast, eco-friendly, conductive, UV protective, anti-odor, antimicrobial, anti-static, breathable; these textiles should also have advanced thermal performance for climate control, durability, improved softness, better tear strength, water/spill repellency, flame retardancy, and wrinkle resistance. However, for use in defense, the demands are much greater and nanotechnology offers numerous advantages. Some nanomaterials that are already being incorporated into textile materials are silver, gold, silicon dioxide, titanium dioxide, zinc oxide, aluminum hydroxides, nanoclay, carbon nanotubes, carbon black, graphene, copper, and iron hydroxides.

Coating with metal and metal oxide nanoparticles is done to impart different properties in textiles. Some successful attempts are mentioned below.

7.5.1 Antimicrobial Fabrics

Microbial contamination is mainly a great concern for textiles. Silver nanoparticles (AgNP) and zinc oxide (ZnO) nanoparticles possess antimicrobial activity naturally, thus coating with AgNPs not only kills microbes but also removes sweat odor, especially when coated on socks or underwear.

7.5.2 Water Repellant (Hydrophobic Fabric), Stain Repellant and Spill Resistant Fabrics

Polyester fibers that are coated with minute silicone filaments are used for making waterproof fabric. For this idea, nature was our teacher, as it has been taken from naturally water-repellant surfaces such as lotus leaves, which have a similar combination of tiny nanostructures and (super)hydrophobic substances. Huang *et al.* [18] have managed to produce superhydrophobic silk by coating it with 70 and 160 nm nanorods of ZnO, simply by immersing silk in a solution of zinc nitrate hexahydrate and then drying it. The Zn coated silk was added to n-octadecanethiol (ODT) solution for 12 hours then dried and baked in an oven. A hierarchical micro-nanoscale topography was observed on the SEM image of the surface of treated silk, which revealed the presence of high-density ZnO nanorods of 70 and 160 nm diameters and an average length of 0.9–1.3 μm. X-ray diffraction results showed that these structures were pure, highly crystalline ZnO, and XPS confirmed that they contained no metallic Zn. The hydrophobic property of both pristine and treated silk was tested. The pristine silk was completely wetted, whereas the water contact angle of the treated silk was 151.93°, making it a superhydrophobic material. To examine the durability of the coating, the fabric was treated with alkaline and acidic solutions, as well as washed.

Schoeller Technologies has already developed a fabric by treating the surface of the fabric with nanospheres, allowing clothing to shed water just like a lotus leaf does. The hilly surface of a nanosphere results in less area with which dirt or water can make contact. Similarly, BASF used the lotus effect to develop a fabric called Mincor® TX TT by packing nanoparticles onto fabric so closely that dust and dirt can't get through to attach themselves to the fabric. Dirt simply stays in a layer of air above the fabric and washes off easily. Now Mincor is being used in awnings, umbrellas and tents.

Nanotechnology has also been applied to create wrinkle and stain resistant clothing. Moreover, plasma creates radical sites on the fabric and can be used to insert nanoparticles onto the surface as well. Stain resistant fabrics can withstand discoloration caused by liquids, oil and grease etc. Drawing inspiration from peaches, Nanotex have made upholstery fabrics which are stain resistant [19]. Nanotex is using nanowhiskers of silica that are tiny hair-like projections which cause liquid to bead up and roll off the surface of fabric.

7.5.3 Self-Cleaning Fabrics

Semiconductors, such as titanium dioxide (TiO2) and zinc oxide (ZnO), are the basic nanoparticles used for photocatalytic self-cleaning fabric, which work as photocatalyst to decompose an organic stain into water and carbon dioxide (CO_2) in the presence of UV light source. The self-cleaning concept is very useful for medical textiles and military uniform fabrics. Additional benefits of such fabric are that they will conserve water and reduce energy, laundry cost and time. Here also, nature has been the source of inspiration for the development of self-cleaning fabric by taking its cue from rice plants, butterfly wings, fish scales, etc. [20, 21].

During photocatalysis, photoreaction is accelerated in the presence of catalyst. Using sunlight, usually ultraviolet light (UV), this process can decompose dirt molecules and organic contaminants can also be degraded into air and water [22].

When the energy is equal to or higher than the band gap, electrons on the photocatalyst surface are excited and escape from the valence band to the conduction band, leading to the formation of electron hole pairs in the surface-excited negatively charged electrons in the conduction band, and the positively charged holes (H+) in the valence band. The created pairs can recombine or get trapped and react with other material absorbed on the photocatalyst. The pairs will cause redox reactions at the surface; the negative electrons (e-) and oxygen will combine to form superoxide radical anions ($O2^-$), whereas the positive electric holes and water will generate hydroxyl radicals (OH^-). Eventually, all the formed highly active oxygen species will oxidize organic compounds to carbon dioxide (CO_2) and water (H_2O). Hence, photocatalyst can decompose common organic matters in the air such as odor molecules, bacteria and viruses [23].

The self-cleaning properties of cotton fabrics coated with nano-TiO_2 and nano-TiO_2 mixed with fumed silica have been shown by Pisitsak *et al.* [24].

Huang *et al.* [18] tested the ZnO nanoparticle coated silk and examined its self-cleaning performance by placing dry methylene blue powder on its surface; and then added water to remove the powder entirely, leaving the treated silk unstained. Moreover, when fabric was treated with alkaline and acidic solutions and then washed, the coating remained durable.

7.5.4 UV Radiation Protection

There have been many successful attempts to confer UV-protecting properties onto different types of textiles, mainly by coating them with nanoparticles of titanium dioxide. UV-protecting performance (i.e., ultraviolet protection factor or UPF) of nanoparticle treated fabrics are usually tested by recording the transmittance of UVA (320–400 nm) and UVB (280–320 nm) through it.

The UV (ultraviolet) protection factor value (UPF) is the ratio of UV radiation measured without the protection of the fabric compared to that with protection. UPF tests are done using a spectroradiometer. A textile for smart fabric should have a minimum of 15 UPF to be rated as UV protective [25]. ZnO-coated fabric in the region of the UVB (280–315 nm) shows excellent UV-blocking ability [26]. Simultaneously, Sivakumar *et al.* [27] showed that cotton fabric coated with larger size nano ZnO and nano TiO_2 with acrylic binder have better UV protection factor (UPF) value than the fabric treated with smaller sized ZnO nanoparticles.

There are many other hazardous ionizing radiations that may be used in war such as gamma rays, X-rays, radioactive particles, either alpha or beta, but unfortunately no smart textile has been developed so far to protect against them. There are some protective suits that has been developed to keep radioactive isotopes out of the body but they cannot shield against the ionizing radiation. The United States Department of Homeland Security defines a Hazmat suit as "an overall garment worn to protect people from hazardous materials or substances, including chemicals, biological agents, or radioactive materials."

7.5.5 Static Resistant or Anti-Static Fabric

Synthetic fabrics, such as polyester and nylon, gather static charge when they come in contact with the head, causing the hair to stand on end. To resist static charge, nanoparticles that conduct electricity, such as zinc oxide, titanium dioxide and antimony-doped tin oxide, are used on fabric to disperse this charge. It is expected that more fabrics made from nanofibers with nanoparticles and monofilaments will be an integral part of the weave to make "smart fabrics."

7.6 Applications of Nanoparticle Coated Smart Textiles

Apart from nanoparticle coated modern textiles, there are next generation smart textiles being developed for use in defense. These smart textiles are developed by fusing them with integrated computing power that can sense and react to external or internal stimuli, communication between devices, conduct electricity or energy, sense, react and protect from hazardous environments. Apart from clothing and apparel, applications of smart textiles have extended to robotics, aerospace, automobiles, medicine, aircraft, surgery and military/defense. On battlefields they virtually can act as life-saving materials.

7.6.1 Healthcare Fabrics

Military forces need smart clothing to increase their safety and effectiveness in hazardous and extreme environmental conditions and to give real-time information to increase protection and survivability. For this it is important to monitor vital signs and treat injuries while simultaneously monitoring environmental hazards, such as toxic gases, and responding rapidly and safely.

Wearable smart textile interfaces integrating sensors, electrodes, communication system and connections realized with conductive and piezoresistive yarns [28, 29], which are based on support from nanotechnology, are areas of focus of the WEALTHY (Wearable Health Care System) and MyHeart projects. Smart textiles are being tried and used in various aspects of healthcare (Table 7.1) for continuous monitoring of injured or sick defense personnel. So that injured army personnel can overcome the problem of a visit to the hospital and get timely treatment, wireless-enabled garments with embedded textile sensors can simultaneously and continuously monitor ECG, respiration, EMG, and other physical activity. Along with monitoring health conditions, the smart textiles are embedded with portable electronic circuit boards for motion assessment, signal pre-processing and Bluetooth connection for data transmission. Wearable sensitized smart garments are composed of conductive fiber grid where sensors are fully integrated.

Collins and Buckley have demonstrated that conductive polymers coated fabric can detect very low ppm limits of hazardous and toxic gases such as ammonia and nitrogen dioxide and chemical warfare simulant dimethyl methylphosphonate (DMMP) [30].

Armed forces deployed in mountains can be tracked if they wear shoes in which Global Positioning Systems (GPS) are incorporated and gloves having heaters or built in LEDs emitting light so that they can be tracked in the dark. Their uniform fabric should maintain the moisture by releasing moisturizer into the skin, maintain and control the body temperature, as well as control the muscles vibration, especially during physical activities, through chromism.

7.6.2 Self-Powered Smart Textiles

Research in the search for a human-like and more human-friendly *smart skin* for robots is based on resistance, capacitance or transistors—all of which need a power supply feed—means higher energy consumption and complex circuits. A Chinese research group lead by Zhang [31] has developed a self-powered smart skin that can solve this problem. In their next report published in 2017, they have combined the triboelectric effect and planar electrostatic induction and applied it in a subtle device structure to create a self-powered analogue smart skin, that can simultaneously detect bending and pressure in one multifunctional device. This sensor is made with two orthogonal CNT-polyurethane sponge strips. The strips are designed in such a way as to detect bending direction and degree simultaneously. The aspect ratio of CNT and porosity of polyurethane is manipulated to improve the sensitivity of this e-skin. Since triboelectric effect has been combined in this fabrication, it will help in distinguishing pressure from bending signal, because pressure generates triboelectric voltage that bending cannot.

This smart skin is intrinsically self-powered. Zhang has explained it as "Triboelectric charges occur in our daily life everywhere when two surfaces touch each other. And when the charged surface approaches a metal block (or electrode) it will induce the opposite charge, which is the so-called 'electrostatic

Table 7.1 Some achievements in smart textiles wearable for healthcare and protection of army personnel.

Item	Produced by
Cotton/Lycra® fabrics are used to make apparel that integrates carbon-loaded elastomer strain sensors and electrodes for electrocardiogram (ECG), electromyogram (EMG) and body movement. To detect ECG and EMG signals, electrodes are knitted using stainless steel-based yarns. To improve contact and match impedance with the skin a hydrogel membrane is applied.	**WEALTHY** (Wearable Health Care System) is an EU funded project
Made wearable textile interfaces integrating sensors, electrodes and connections using conductive and piezoresistive yarns.	**MyHeart** is another EU funded project
Smart Shirts have come on the market for similar applications mentioned above.	Sensatex™
Another smart shirt product is the Life Shirt® system that offers continuous ambulatory monitoring systems.	VivoMetrics®
Smart glove with sensors (noninvasive sensors) that can monitor emotional, sensory, and cognitive activities, for detection of the activity of the autonomic nervous system, which is responsible for the body's involuntary vital functions and to measure skin temperature, skin electrical conductance, and skin potential. For this a micro-sensor (0.45 mm) is integrated in the glove to monitor skin temperature and electrodes measure the skin's electrical activity.	**MARSIAN** (Modular Autonomous Recorder System for the measurement of Autonomic Nervous) system

(*Continued*)

Table 7.1 Some achievements in smart textiles wearable for healthcare and protection of army personnel. (*Continued*)

Item	Produced by
Two approaches (i) electrode integration to embroider commercially available silver/silver chloride electrodes into hairnet glove and (ii) a 3D structure made of Kapton® copper foil (150 mm thick) with electrodes covered in silver; have been successfully attempted.	
Advanced e-textiles for firefighters and fire victims are made to perform on body biochemical sensing within a textile. These fabrics monitor the surrounding environment to detect any potential risks. The project is to add a wearable sensing and transmission systems to monitor health, activity, position, and environment, with information relayed both to the individual and also to a central monitoring unit.	**ProeTEX** is another EU supported project
Making fabrics to monitor the wearer's health.	Biotex
A SmartShirt™ System is developed that in case signals are below significant values, a wireless communication system automatically can send a distress signal to dial telephones or pager messages, thus offering healthcare and safety by monitoring exact location.	Sensatex
Aquapel™ is water repellent, moisture wicking fabric which gives cool comfort.	**Nanotex** is supplied by Crypton Inc.

induction' effect." Moreover, it needs only 4 electrodes to have mm 2D resolution. Thus, it assisted in reducing the complexity of the signal processing circuits.

7.6.3 CNT-Based Smart Fabrics

CNT-based smart fabrics (CNFs) have high mechanical strength, extraordinary structural flexibility, high thermal and electrical conductivities, novel corrosion and oxidation resistivities, and high surface area, which makes them a very promising candidate for next-generation smart textiles and wearable devices.

One of the needs of smart textiles is real-time tracking of the location and vital signs of the wearer, which is important for not only the military but also firefighters and police. The challenge in making smart fabric is achieving a seamless and invisible integration of required electronics into the fabric. This demands fibers with desired electronic functionality without compromising strength and comfort, as it needs complicated inputs, like implementing electronic circuits and communication system, particularly storage and antenna system for transmission of data. CNTs (carbon nanotubes) are highly conductive, seamless cylindrical hollow tubes made of a sheet of folded graphene having superior electronic and mechanical properties, and also exhibit high-frequency properties in the gigahertz range [32]. CNTs are being incorporated in smart fabric as they can be embedded in many objects that are operated at high frequencies and are suitable for a range of communication and electronic applications. CNTs have an electric current-carrying capacity of 1000 times higher than copper wire and are stronger than steel.

CNT-based polymer fabrics can be used to create invisible antenna patches incorporated in textile. According to Foroughi *et al.* [33], a typical patch of 20 antenna elements at 60 GHz will be around 50 mm × 5 mm and could be used to facilitate low-power and high-speed communication at any site. The antenna array needs to be incorporated in the garment because it will create a communication link between sensors, and on body devices like drug delivery nanodevices injected into the blood or sensors incorporated in e-skin.

CNTs have shown their application in supercapacitors, actuators, sensors, lightweight electromagnetic shields, nanoelectronics devices, probe tips for AFM, and in aerospace industries for dissipation of electrostatic charges [34–36].

7.6.3.1 CNT and Metallic Antennas for Smart Textiles

Bluetooth has had major appeal in the development of on-body devices in smart fabrics by integrating transmitters and receivers into garments, uniforms and headgear. There are already radio-frequency identification (RFID) tags for subdermal implants. Now they are focusing on high frequency range, i.e., gigahertz range. This does not deny the need for lower frequency range, where the size of antenna is larger and can be used for location tracking as the signal can penetrate high density structures like rocks, etc. Trials are on to energize garments through wireless signal with high powered transmitter at a long distance. Traditionally, metals are used for antenna. But to integrate in garments we need more flexible, corrosion-free material. Hanson [37] suggested CNTs are metallic or semiconducting depending on geometry. But simultaneously, Attiya [38] showed low frequency limit of CNT antenna. He suggested that CNT communication might be limited to short-range applications. However, Sharma *et al.* [39] reported optical *Rectenna* (an antenna that operates at optical frequency) using CNT sheet as the receiving antenna connected to photodiode (switching at a speed of 1 PHz). Another approach that shows promise suggested by Zahir *et al.* [40] and patented by Kelkar and Zaghloul [41] is the use of *patch antenna* (it is a flat piece of material that allows energy to radiate out from its surface). Many trials using CNTs are going on in the development of CNT-based antennas for low frequency application. It is plausible to get suitable forms and composition of CNT-based material for smart textiles or electronic textiles.

7.6.3.2 Cotton Coated with MWCNT for Energy Storage

Bharath *et al.* [42] have developed a MWCNT coated cotton fabric by simply dipping the fabric many times into a surfactant, sodium dodecyl sulphate (SDS), with MWCNT dispersed in it. Further treatment with KOH helped in loading of MWCNT through hydrogen bonding with glycosidic group present in cotton (cellulose) fibers. The loaded MWCNT was assessed by optical absorption spectra (λmax = 442 nm) and FE-SEM. The interaction of MWCNT with cotton fibers enhanced its electronic properties. This flexible conducting cotton fabric had a resistance of <1.5 k cm^{-2} and capacitance of 40 μF. The best part is that the resistance of MWCNT-coated cotton fabric can be easily tuned by controlling the amount of MWCNT coating and chemical (with 5% HNO_3) treatment to improve the functionality of loaded MWCNT. Moreover, a capacitor formed by placing two MWCNT-coated fabrics between etched PCB plates (terminal contacts) shows the charging capacity of about 1 F. Such lightweight, flexible cotton fabrics are environmentally friendly and low cost and could be used in embedded healthcare and wearable electronic systems.

7.6.3.3 CNT Braided Fabric for Monitoring Composites

A braided CNT fabric for monitoring composites is being developed to have built-in functionality to continually monitor and diagnose their own health states. Luo *et al.* [43] have developed an *in-situ* monitoring of the processing stages of composites by co-braiding CNT-enabled fiber sensors into the reinforcing fiber fabrics. These CNT-enabled smart fabrics are highly sensitive to monitor and quantify various events of composite processing, including resin infusion, onset of crosslinking, gel time, degree and rate of curing. They have suggested that the smart fabrics that are readily and noninvasively integrated into composites can provide life-long structural health monitoring of the composites, including detection of deformations and cracks. The authors have demonstrated the robust and versatile sensory technology of smart fabric for diagnosing and evaluating the health states of polymeric composites from the manufacturing process to the service stage and finally to failure. And by co-braiding MWCNT enabled fiber rovings into a fiberglass woven preform, the use of the smart fabric sensor to provide *in situ* resin infusion and curing information during the vacuum assisted resin transfer molding (VARTM) process of composite manufacturing has been demonstrated. The smart fabric sensor readily and noninvasively integrated into the laminate proved to be desirable for monitoring the strain and stress states, as well as detecting the failures of the host structure.

7.6.3.4 CNT-Based Smart Electronic Textile

One of the needs of smart and wearable devices is fiber electronics CNT-based prototypes of fiber electronics (memristors and fiber superconductors), which have been demonstrated by many researchers.

In 2010, Jeong *et al.* [44] showed that sandwiching graphene oxide between two planar electrodes shows a memory effect. Sun *et al.* [45] have developed a fiber-based memristor by cross-stacking two CNT fibers coated by graphene oxide nanosheets. This fiber memristor indicated a nonvolatile behavior as the device showed an abrupt current increase when the positively weeping voltage exceeded 3.5 V, which suggests a state switch from OFF (high-resistance state) to ON (low-resistance state). When the voltage continuously swept back to

0 V, the device remained in the OFF state. Moreover, the device showed a write-once-read-many-times (WORM) behavior. Also, the ON/OFF ratio was about 1000, which is comparable to the values of graphene-oxide-based memristors using metal electrodes; the ON/OFF ratio was increased to 109 by simply adjusting the concentration and acidity of the graphene oxide aqueous solution used for the coating by Li *et al.* [46].

In another attempt, Bykova *et al.* [47] created a CNT–MgB_2 superconducting fiber having critical temperature of 37.8 K and a gravimetric critical current density 10 times higher than that of dense MgB_2 wires. This composite is porous in nature and has low density (0.124 g cm^{-3}) and is flexible, weavable and knittable. Its porosity may enable fast cooling.

7.7 Sensors for Smart Textiles

Besides the above-mentioned fiber actuators, sensors, data processing unit, communication system, interconnections and energy supply harvesting and storage devices are also an integral part of wearable smart textiles. A sensor is a device which detects or measures a physical property (light, heat, motion, moisture, pressure, or many other environmental phenomena) and records, indicates, or responds to it. Components of a sensor vary with application. For example:

(i) Digital sensors within an instrument have a microcontroller for signal timing and data storage;
(ii) Analog sensors are analyzed by a PC; and
(iii) Multiple sensors are displayed over the internet.

Conventional transducers are large, but generally reliable, based on the older technology of thermocouple (for temperature difference) and compass (magnetic) for direction.

Whereas, *microelectronic sensors* are millimeter sized, highly sensitive and less robust and can be (a) Photodiode/phototransistor can detect photon energy (light), infrared, as proximity/intrusion alarms; (b) Piezoresistive pressure sensor can measure air/fluid pressure; (c) Microaccelerometers can detect vibration, Δ-velocity during a vehicle crash; (d) Chemical sensors can detect O_2, CO_2, Cl, nitrates (explosives); and (e) DNA sensor arrays can match DNA sequences.

Though there have been many technical challenges, some wearable e-textiles where sensors, actuators, and production techniques were used have been developed to seamlessly embed electronic features into traditional wearable textiles. However, efforts are underway to adapt them for use in smart fabric/e-textile fabricated by various methods such as embroidery, sewing, weaving, nonwoven, knitting. spinning, breading, coating and printing. The conductivity changes in response to external pressure/deformation have been exploited to produce textile-based mechanical sensors.

7.7.1 Temperature Sensor

Temperature sensors are made using fabrics that are flexible, e.g., plastics and polyimide sheets. Sensors are attached or integrated on to the temperature sensitive textile fabric. The resistance temperature detectors (RTDs) are composed of platinum/nichrome (NiCr) coated on flexible surfaces. Its resistance changes linearly with temperature. All the conductive polymers and carbon-based conductive particle polymers give temperature-dependent response. They show a decrease in resistance at high temperatures [48]. Fiber optic sensors can also be used to sense temperature changes.

Sensors in textiles are needed for various applications, e.g., for temperature. Diaconescu *et al.* [49] have developed electronic smart textiles containing integrated electronic device (sensor) or transducer) for measurement and display of temperatures

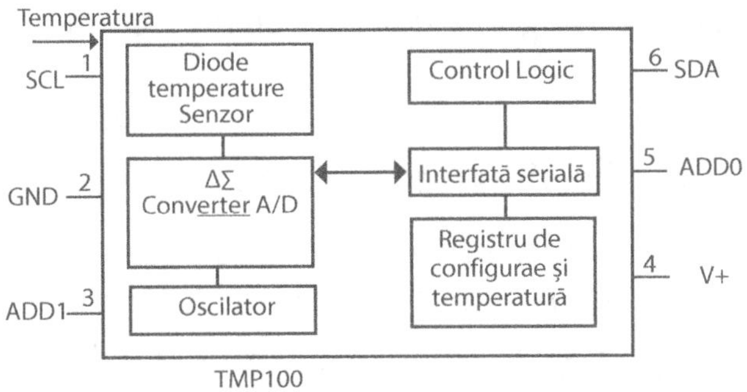

Figure 7.9 TMP100 sensor's internal structure. (Source: TMP100, TMP101 – Texas Instruments: www.ti.cpm/lit/ds/pdf)

inside and outside, being used as commercial sensor model TMP 100. Apart from monitoring, it can also control the temperature. It has a wiring diagram design for the system; key (code) writing and implementation for the measured temperature display with digital sensor. The transducers include a primary transmitter, also known as sensitive element or sensor. The sensor can directly retrieve information about the desired parameter from the process.

The inner structure of the temperature sensor as suggested by Texas Instruments is shown in Figure 7.9.

7.7.2 Humidity Sensitive Textiles

Humidity textile sensors are based on resistance and capacitance, i.e., the resistive humidity textile sensors respond to moisture variation by changing its conductivity, while the capacitive humidity textile sensors answer to water vapor by varying its dielectric constant.

Polymers suitable for capacitive humidity sensors include polyethersulfone (PES), polysulfone (PSF), and divinyl siloxane benzocyclobutene (BCB). These sensitized substrates (polymers) can be woven into textiles. Flexible transistors are also humidity sensing devices that change conductivity with the humidity levels.

7.7.3 Capacitive Pressure Sensors

Capacitive fibers are fabricated using silicon fiber sputtered with metals [50]. Organic polymers provide help in overcoming the stiffness of silicon crystals. Capacitive pressure sensors are sewn, snapped, or glued onto a fabric substrate. Sensor is then welded to other electronics or wires. Compliant conductive materials used for making capacitors can also act as conductive plates separated by dielectrics. An attempt has been made to attach these conductive plates to the fabric by different methods such as (i) weaving [51], (ii) sewing [52], (iii) embroidered on conductive thread/fabrics, (iv) painting, printing, sputtering or screening with conductive inks [53] or conductive polymers [54]. Usually the dielectrics used are synthetic foams, fabric spacers, and/or soft nonconductive polymers. The capacitance of a capacitive pressure sensor depends on the area of two conductive parallel plates, the conductive material and the distance between each other.

Metal, optical fibers and conductive polymers are integrated into the textile structure, thus supplying electrical conductivity, sensing capabilities and data transmission. Crosslite™ is producing capacitive pressure sensors with higher resolutions that can be used to sense pressures over time.

7.7.4 Resistive Pressure Sensors

To measure pressure, these resistive pressure sensors are made of different conductive materials in different structures. There is a correlation between pressure and electrical resistance. The resistive materials are sewn, embroidered or glued to the textile as per need. In a resistive pressure sensor, an electric resistance increases when the resistive material is stretched or compressed. According to Ohm's Law (V = R*I), for the same electric current, a higher resistance makes the output voltage increase. Therefore, the stretch or compression can be correlated to the sensed voltage [55]. The conductive material and production technique influence the sensitivity and sensed pressure range. Hence, another way of fabricating pressure sensitive fabrics is to coat them with conductive silicone.

7.7.5 Optical Textile Sensors

The optical textile sensor is based on the photosensitivity in optical fibers that can sense variation of the light intensity or the amplitude by a Fiber Bragg Grating (FBG) sensor. It was first developed by the Hill group when the photosensitivity in optical fibers was discovered [56]. Glass optical fibers of micron diameters (in the microns range) are suitable for seamless textile integration. For this the optical fiber light source used is a small light emission diode (LED), and the light amplitude at the end of the optical fiber can be sensed with a small photodetector. The light amplitude will change with the movement of textile, allowing the sensing of textile displacements and pressures. The optical textile sensors are useful when the electrical currents cannot cross textile substrates. When the elastic fabric is stretched, the light amplitude passing through the fiber increases, which increases the output voltage coming out the photodetector.

7.8 Actuators for Smart Textile

Inherently conducting polymers (ICPs) are used for kinesthetic and physiological monitoring, and for patient rehabilitation [57]. PANi and PPy (the electroactive polymers) are used as sensing devices and may also be configured as actuator by configuring them as electrode within an electrochemical cell;

to which, when potential is applied, the ICP electrode changes its dimension and works as a mechanical actuator. Such actuators are integrated within textiles, which enable fabrics to have motor functions. Scientists are working on using this unique property to develop artificial muscles.

The actuation property of ICPs is due to the volume change of ICPs. Application of positive potential leads is used to remove electrons from the polymer backbone and incorporation of dopant ions to maintain electrical neutrality.

According Della Santa *et al.* [58], ICP-based mechanical actuators can achieve average stresses ~10 to 20 times than those generated from natural muscle. Hara *et al.* [59] found that the realized strains (>20%) were comparable to natural muscle. Later, Wu *et al.* showed that fast freestanding beam actuation was achieved with an operational frequency of up to 40 Hz.

Wu *et al.* [60] reported more than one million redox cycles using an ionic liquid 1-butyl-3-methylimidazolium tetrafluoroborate (BMI-BF4)/PANi fiber actuator system, having minimal decrease in actuation strain.

CNT and polymer (PANi) composite of ICP have exhibited increased electronic conductivity of up to ~750 S cm^{-1} and chemical properties, and an increase in tensile strength and elastic modulus of 50% to 120%. CNTs help in enhancement of high strength, robustness, good conductivity, and pronounced electroactivity. Thus, they are potentially useful in electronic textile applications, such as for enhanced force generation when incorporated into fabric as an actuator and the improved conduction when used as the connection wire. Films of ICP nanofibers have been used in sensors to detect chemical vapors that interact with ICPs and change their conductivity.

These smart fabrics help in tracking control of shape-memory-alloy (SMA) actuators.

7.9 Summary

Nanoparticles are incorporated in textiles to increase their smartness. Nanoparticles are coated on fabrics to improve the performance and functionality of textiles. Nanoparticle coated fabrics are highly durable fabrics. Coating with nanoparticles has other advantages such as providing antibacterial, water-repelling, spill resisting, stain repelling/releasing, UV protecting and self-cleaning apparel. Moreover, it maintains the breathability and tactile properties and the textile can resist static charge.

Nanomaterials integrating technologies have enabled passive, noninvasive sensing of wearers and their environs. Moreover, incorporation of nanoparticles has solved the problem of interconnecting wearable computing. At present the conventional interconnections are done with silicon and metal components, which are highly incompatible with the soft textile substrate. Integration of nanoscale technology preserves the tactile and mechanical properties of the fabric, keeping smart textiles flexible enough to be comfortably worn for long periods of time. The area of smart textiles is highly specialized and involves input from materials scientists, sensor technologists, engineers, wireless networking, and computer experts.

References

1. Marvin, C., When Old Technologies Were New: Thinking About Electric Communication in the Late Nineteenth Century, Oxford University Press, USA, 1990.
2. Khare, R. and Pandey, J., Smriti. *Sci. Eng. Appl.*, 1, 4, 22–26, 2016.
3. Ye, L., Fei, F., Wang, T., *China: Five Thousand Years of History and Civilization*, p. 410, City University of Hong Kong Press, Hong Kong, ISBN: 978-962-937-140-1, 2007.

4. Gere, C. and Rudoe, J., *Jewellery in the Age of Queen Victoria: A Mirror to the World*, British Museum Press, London, UK, 2010.
5. Wainwright, H.L., Design, evaluation, and applications of electronic textiles. In Performance Testing of Textiles on Science Direct Performance Testing of Textiles Methods, Technology and Applications Woodhead Publishing Cambridge, England, 2016.
6. Gregory, R.V., Samuel, R.J., Hanks, T., National Textile Centre Annual Report, USA, 2010.
7. Coyle, S., Wu, Y., Lau, K.T., De Rossi, D., Wallace, G., *MRS Bull.*, 32, 434–432, 2007.
8. Oakes, J., Batchelor, S.N., Dixon, S., *Coloration Technol.*, 12, 237–244, 2005.
9. Chiang, C.K., Fincher, C.R., Jr., Park, Y.W., Heeger, A.J., Shirakawa, H., Louis, E.J., Gau, S.C., MacDiarmid, A.G., *Phys. Rev. Lett.*, 39, 1098, 1977.
10. Baughman, R.H., *Synth. Met.*, 78, 339, 1996.
11. Huang, S.V., Weiller, B.H., Kaner, B.H., *J. Am. Chem. Soc.*, 125, 314, 2003.
12. Wallace, G.G., Teasdale, P.R., Spinks, G.M., Kane-Maguire, L.A.P., *Conductive Electroactive Polymers: Intelligent Materials Systems*, 2nd ed, CRC Press, Boca Raton, FL, 2002.
13. Smith, P.J., Catalog of the exhibition "Body Covering" held at the Museum of Contemporary Crafts, (April 6 through June 9), 1958. Publisher, American Craftsmen's Council. New York City.
14. Knobloch, A., Manuelli, A., Bernds, A., Clemens, W., *J. Appl. Phys.*, 96, 2286, 2004.
15. Chen, Q.L., Xue, K.H., Shen, W., Tao, F.F., Yin, S.Y., Xu, W., *Electrochim Acta.*, 49, 24, 4157–4161, 2004.
16. Mather, R.R. and Wilson, J., *Intelligent Textiles and Clothing*, H.R. Mattila (Ed.) pg. 206–216, CRC Press Boca Raton Boston New York Washington, DC, 2006.
17. Runnerstrom, E.L., Llordés, A., Lounis, S.D., Milliron, D.J., *Chem. Commun.*, 50, 10555–10572, 2014, doi: 10.1039/C4CC03109A.

18. Huang, J., Yang, Y., Yang, L., Bu, Y., Tian, X., Gu, S., Yang, H., Ye, D., Xu, W., *Mater. Lett.*, 237, 149–151, 2019.
19. Boysen, E., Muir, N.C., Dudley, D., Peterson, C., Nanotechnology makes fabric water and stain resistant, in: *Nanotechnology for Dummies*, 2nd ed., ISBN: 978-0-470-89191-9, 2011.
20. Barthlott, W. and Neinhuis, C., *Planta*, 202, 1–8, 1997.
21. Bixler, G.D. and Bhushan, B., *Soft Matter*, 8, 11271–11284, 2012.
22. Kumar, B.S., *IOSR-JPTE*, 2, 1, 2348–0181, 2015.
23. Saad, S.R., Mahmed, N., Abdullah, Mohd Mustafa Al Bakri, Sandu, A.V., IOP Conf. Ser.: Mater. *Sci. Eng.*, 133, 012028, 2016.
24. Pisitsak, P., Samootsoot, A., Chokpanich, N., *KKU Res. J.*, 18, 2, 200–211, 2013.
25. Senic, Z., Bauk, S., Vitorovic-Todorovic, M., Pajic, N., Samolov, A., Rajic, D., *Sci. Tech. Rev.*, 61, 3-4, 63–72, 2011.
26. Shateri-Khalilabad, M. and Yazdanshenas, M.E., *Text. Res. J.*, 83, 10, 993 1004, 2013.
27. Sivakumar, A., Murugan, R., Sundaresan, K., Periyasamy, S., *IJFTR*, 38, 285–292, 2013.
28. Paradiso, R., Loriga, G., Taccini, N., *IEEE Trans. Inf. Technol. Biomed.*, 9, 337–44, 2005.
29. Paradiso, R., Belloc, C., Loriga, G., Taccini, M., *Personalised Health Management Systems*, C. Nugent, P.J. McCullagh, E.,.A.E.A. McAdamsLymberis (Eds.), pp. 9–16, IOS Press, Amsterdam, 2005.
30. Collins, G.E. and Buckley, L.J., *Synth. Met.*, 78, 2, 93–101, 1996.
31. Shi, M., Zhang, J., Chen, H., Han, M., Shankaregowda, S.A., Su, Z., Meng, B., Cheng, X., Zhang, H., *ACS Nano*, 10, 4, 4083–4091, 2016, doi: 10.1021/acsnano.5b07074.
32. Kshirsagar, D.E., Puri, V., Maheshwar, Sharon, Madhuri, Sharon., *Carbon Sci.*, 7, 4, 245–248, 2006.
33. Foroughi, J., Mitew, T., Ogunbona, P., Raad, R., Safaei, F., *IEEE Consum. Electr. M.*, 5, 4, 105–111, 2016, doi: 10.1109/MCE.2016.2590220.
34. Baughmann, R.G., Cui, C., Zakhidov, A.A., Oqbal, Z., Barisci, J.N., Sinks, G.M., Wallace, G.G., Mazzoldi, A., De Rossi, D., Rinzler, A.G., *Science*, 284, 5418, 1340–1344, 1999.
35. Ferrer-Anglada, N., Kaempgen, M., Roth, S., *Phys. Status Solidi B*, 243, 13, 3519–3512, 2006.

36. Bokobze, L., *Polymer*, 47, 17, 4907–4920, 2007.
37. Hanson, G.W., Fundamental transmitting properties of carbon nanotube antennas. *IEEE Trans. Antennas. Propag.*, 53, 11, 3426–3435, 2005.
38. Attiya, A.M., *Prog. Electromagn. Res.*, 94, 419–433, 2009.
39. Sharma, A., Singh, V., Bougher, T.L., Cola, B.A., *Nat. Technol.*, 10, 12, 1027–1032, 2015.
40. Zahir, H., Wojkiewicz, L., Alexander, P., Kone, L., Belkacem, B., Bergheul, S., Lasri, T., *IET Microw. Antennas P.*, 10, 1, 8893, 2016.
41. Keller, S.D., Zaghloul, A., Shanov, V, *IEEE Trans. Antennas Propag.*, 62, 1, 48–55, doi: 10.1109/TAP.2013.2287272, 2014.
42. Bharath, S.P., Manjannaj, J., Javeed, A., Yallappa, S., *Bull. Mater. Sci.*, 38, 1, 169–172, 2015.
43. Luo, S., Wang, Y., Wang, G., Wang, K., Wang, Z., Zhang, C., Wang, B., Luo, Y., Li, L., Liu, T., *Sci Rep*, 7, 44056, 2017, doi: 10.1038/srep44056.
44. Jeong, H.Y., Kim, J.Y., Kim, J.W., Hwang, J.O., Kim, J.E., Lee, J.Y., Yoon, T.H., Cho, B.J., Kim, S.O., Ruoff, R.S., Choi, S.Y., *Nano Lett.*, 10, 11, 4381–6, 2010.
45. Sun, G., Liu, J., Zheng, L., Huang, W., Zhang, H., *Angew. Chem. Int. Ed.*, 52, 50, 13351–5, 2013.
46. Li, R., Sun, R., Sun, Y., Gao, P., Zhang, Y., Zeng, Z., Li, Q., *Phys. Chem. Chem. Phys.*, 17, 7104, 2015.
47. Bykova, J.S., Lima, M.D., Haines, C.S., Tolly, D., Salamon, M.B., Baughman, R.H., Zakhidov, A.A., *Adv. Mater.*, 26, 7510–7515, 2014.
48. Lay, M., Pèlach, M.À., Pellicer, N., Tarrés, J.A., Bun, K.N., Vilaseca, F., *Carbohydr. Polym.ss*, 165, 86–95, 2017.
49. Diaconescu, V.D., Vornicu-Albu, L., Diaconescu, R., *Buletinul AGIR nr*, 3, 31–35, 2015.
50. Shen, L., Healy, N., Xu, L., Cheng, H., Day, T., Price, J., Badding, J., Peacock, A., *Opt. Lett.*, 39, 5721–5724, 2014.
51. Bhattacharya, R., Van Pieterson, L., Van Os, K., *IEEE T. Comp. Pack. Man.*, 2, 165–168, 2012.
52. Avloni, J., Lau, R., Ouyang, M., Florio, L., Henn, A., Sparavigna, A., *J. Ind. Text.*, 38, 55–68, 2008.

53. Komolafe, A., Torah, R., Tudor, J., Beeby, S., *Proc. Eurosensors*, 1, 4, 613, 2017.
54. Yip, M.C. and Niemeyer, G., *Proceedings of the 2015 IEEE International Conference on Robotics and Automation (ICRA)*, pp. 2313–2318, 2015.
55. Stewart, R. and Skach, S., Initial investigations into characterizing diy e-textile stretch sensors. *Proceedings of the 4th International Conference on Movement Computing*, London, UK, 28–30 June 2017, ACM: New York, NY, USA, 2017.
56. Hill, K., Malo, B., Bilodeau, F., Johnson, D., *Annu. Rev. Mater. Sci.*, 23, 125–157, 1993.
57. Tognetti, A., Lorussi, F., Bartalesi, R., Quaglini, S., Tesconi, M., Zupone, G., De Rossil, D., *J. Neuroeng. Rehabil.*, 2, 8, 2005.
58. Della Santa, A., Rossi, D.D., Mazzoldi, A., *Smart Mater. Struct.*, 6, 23, 1997.
59. Hara, S., Zama, T., Takashima, W., Kaneto, K., *Synth. Met.*, 146, 1, 47–55, 2004.
60. Wu, Y., Alici, G., Spinks, G.M., Wallace, G.G., *Synth. Met.*, 156, 1017, 2006.

8

Nanomaterial for Adaptive Camouflage and Structure

Angelica Sylvestris Lopez Rodriguez

Chemical Engineering Department, Juárez Autonomous University of Tabasco, Mexico

When grand plans for scientific and defence technologies are made, do the people in power think about the sacrifices the people in the laboratories and fields have to make?

A. P. J. Abdul Kalam (Missile & Nuclear Weapon Scientist)
Ex-President of India

8.1 Introduction

The term "nanomaterials" includes all those materials developed with at least one dimension in the nanometric scale. In this area, atomic aggregates (clusters) and particles up to 100 nm in diameter, fibers with diameters less than 100 nm, and thin sheets of thickness less than 100 nm are included.

Nanomaterials exist and are used in practically all socio-economic sectors, such as health, energy, textiles, communication and information technologies, security, transport, etc. They include the manufacture of nanostructured materials, nanoparticles, nanopowders, nanoporous materials, nanofibers,

Madhuri Sharon, et al. Nanotechnology in the Defense Industry: Advances, Innovation, and Practical Applications, (275–288) © 2019 Scrivener Publishing LLC

fullerenes, carbon nanotubes, nanowires, dendrimers, molecular electronics, quantum dots, thin films, and others.

The incorporation of functional nanomaterials to textile fibers is giving rise to a new generation of functional fibers with the capacity to respond to external impulse with new properties.

When matter is manipulated at the small scale of atoms and molecules, it demonstrates entirely new properties. Therefore, scientists use nanotechnology to create innovative, low-cost materials, devices and systems with unique properties. Nanotechnology is the synthesis and application of ideas from science and engineering towards the understanding and production of novel materials and devices at the nanoscale.

Nanoscale materials exhibit properties very different compared with macroscale materials, enabling unique applications. For example, opaque substances become transparent (copper); inert materials are transformed into catalysts (platinum); stable materials are transformed into fuels (aluminum); solids become liquid at room temperature (gold); insulators become conductive (silicone). Materials such as gold, which is chemically inert at normal scales, can serve as catalysts at nanoscale.

There are three basic categories of nanomaterials from the commercial and development point of view: metal oxides, nanoclay and carbon nanotubes. The most advanced from the commercial point of view are the metal oxide nanoparticles. Although the science of nanotechnology is relatively new, the existence of devices and functional structures of nanometric dimensions is not; furthermore, such structures have existed on Earth from the origin of life [1].

Nanotechnology has extendable applications; the manufacturing of nanomaterials is one of the most important criteria in the industry area. Nanotechnology could have the potential to solve many of humanity's problems if it is responsibly developed. With the help of nanotechnology, the following can be achieved:

- Fabricate new materials such as clothes that change color (camouflage);
- Synthesize new adhesives and new materials constructed with self-cleaning ability;
- For medical advances, including the administration of medicines, the detection and treatment of diseases such as cancer. With nanotechnology one can build small "blood vessels" that transport medicine directly to the tumor of cancer to destroy it. Further benefits can be achieved in the future [2].

8.2 Camouflage

The word camouflage has its origin in the French word "camoufler" which means "to disguise." Camouflage is the ability of some living beings to go unnoticed, by coloring or the appearance of their body, imitating the appearance of some inanimate object of the surrounding environment or some other living being. Camouflage refers to the action of disguising the presence of weapons, troops, war material, ships, etc., giving an appearance that can deceive the enemy. In the military environment, camouflage is fundamental, it started during the First World War with the painting of American gray ships, confusing the lookouts of the German submarines. Later, the camouflage was transferred to the airplanes, where its light gray color made them blend with the cloudy depths of the sea.

Uniforms also followed this trend, the most commonly used camouflage being a mixture of brown and green in different shades. This camouflage tried to copy the pattern of a leafy forest with different colors and shadows, formed by branches, leaves, trunks, among others. The U.S. Army changed this type of camouflage to brown and black in Somalia, adapting to a desert environment. Subsequently, all uniforms changed from six to three colors, to be more effective camouflage with the environment.

The modern definition of camouflage may be given as "delay or deny detection of a military target by detectors operating over multispectral wavelength region of electromagnetic spectrum or non-electromagnetic radiation, e.g., acoustic, magnetic, etc." The concept of camouflage is as old as nature, and it has been an integral part of it. All animals, small or big, are found to employ several methods of concealment and disguise for self-preservation, both in defense and offence. One example is a creature called the European Buff-tip moth, which disguises itself as a broken twig on a tree [3]. Other well-known examples in nature include the chameleon, which has the ability to change its color to blend itself into the surrounding environment.

There are three types of camouflage in nature: (i) *Protective mimicry,* which provides the ability to camouflage the animals that own it (stick insects, some moths, frogs, flounders, lizards, etc.); (ii) *aggressive mimicry,* which, contrary to the previous one, serves to surprise and attack the prey of the animals that possess it (felines, shark, chameleon, etc.); and (iii) *Batesian mimicry*, which is adopted by generally harmless animals to obtain the appearance of another poisonous or dangerous species for his predator (moths, snakes, octopuses, etc.).

Camouflage has attracted more and more interest due to its special properties such as making animals or objects invisible in the environment. Many researchers have done some excellent work in this field [4–10].

Camouflage is used worldwide, principally in the army. The usual green-gray camouflage allows hiding in foliage during the day, but at night a soldier is vulnerable to infrared sensors. The applications of intelligent structures can be found in practically all fields, such as aerospace, biomedicine, etc. [11].

8.3 Camouflage for the Military

Military camouflage is the skill by which a target goes unnoticed to the eyes of an enemy soldier or vehicle, by blending with the environment that surrounds it. The uniform can also mimic natural or artificial brush, such as mud, snow, branches, etc.

The first camouflage was done around 1915 by the French army [12]. Years later, in World War II, the original patterns and techniques were much more developed, and the color mixtures were better achieved, as well there being more widespread use of this form of concealment. Thanks to modern technology, the use of modern materials, research and the need to adapt changes in environments and war situations, there are a wide variety of camouflage that are developed for the best performance for the environment.

During the First World War, several war camouflage techniques were tried. One was cubism, used by the British and Americans to protect ships from the threat of the new U-Boat German submarines. The allies, being demoralized and without effective defenses, used aerial scouts and photography. The French were the pioneers in hiring artists to devise methods that would diminish the perception of their troops, as well as elements and equipment of destruction. Most of the attempts failed, but it continued to be perfected.

The uniforms abandoned bright colors and moved on to khaki, which was much more suitable for going unnoticed on the ground. In addition, the soldiers worried about preparing their own equipment, painting their helmets or dying their clothes. Sometimes, on fresh paint, they used sand to improve the final result and achieve a matte finish. This was fundamental because any brightness or reflection of sunlight uncovered their positions. Each type is oriented to certain scenarios to help as much as possible to blend in with the environment in order to have more chances of success. Some of the different types of camouflage used by different armies for their protection are mentioned in the following section.

8.4 Types of Camouflage

8.4.1 Woodland Camouflage

This is probably the most popular and most used camouflage and a type of camouflage that everyone knows. It is the classic mixture of brown and green in different shades (Figure 8.1). This camouflage tries to copy the pattern of a leafy forest with different colors and shadows formed by branches, leaves, trunks, etc. Woodland camouflage has been used for many years but research and experience has taught that it is not the best type of camouflage for use in all circumstances.

8.4.2 Desert Camouflage

The six-color desert camouflage, nicknamed by American soldiers as "chocolate chip camo," was first used in Somalia and more widespread in Operation Desert Storm in Iraq.

The typical woodland camouflage did not work in the sandy environment of deserts and the adobe color of the houses in Iraq and Somalia. Therefore, this type of camouflage was invented with shades of chestnut, brown, and black, which appear in a desert environment.

The type of material used in the uniforms also changed to a mixture of polyester and cotton to withstand high temperatures and facilitate perspiration.

8.4.3 Desert Camouflage of Three Colors

In the Middle East, soldiers have changed their 6-color uniforms to 3-color uniforms that is even better for the environment in which their operations take place. The camouflage of three colors contains chestnut, beige and brown (Figure 8.1), allowing the soldier to camouflage himself almost perfectly in desert environments, for example Iraq. The type of material used in the uniforms also changed to a mixture of polyester and cotton to withstand high temperatures and facilitate perspiration.

8.4.4 Digital Army Combat Uniform (ACU) Camouflage

To imagine digital ACU, think about the Tetris game with its falling blocks and the pattern they form with the different mixed colors. Well, the green and brown tones have been mixed in this type of camouflage as if they were pixels of a digital camera, creating a very universal and multifunctional type of camouflage that adapts well to different environments and situations (Figure 8.1).

This type of camouflage adapts well to wooded environments not too dense with clearings, the desert, and also in urban environments. It is not the perfect solution but the best available for very varied environments where the soldier does not have the possibility to change his uniform according to mission and environment because the missions are developed at the same time in different environments. This is one of the best uniforms to use in Spain and other countries in the Mediterranean where there are many different types of environments and where suddenly the forest changes to a ravine, a village or a sandy area.

8.4.5 Tiger Stripe Camouflage

The best camouflage for the jungle or very leafy forests is undoubtedly tiger stripe camouflage. It was invented for use in Vietnam and now it has turned out to be the best available camouflage for environments where the dark shades, yellow of the dim light and the dark green alternate. This camouflage looks

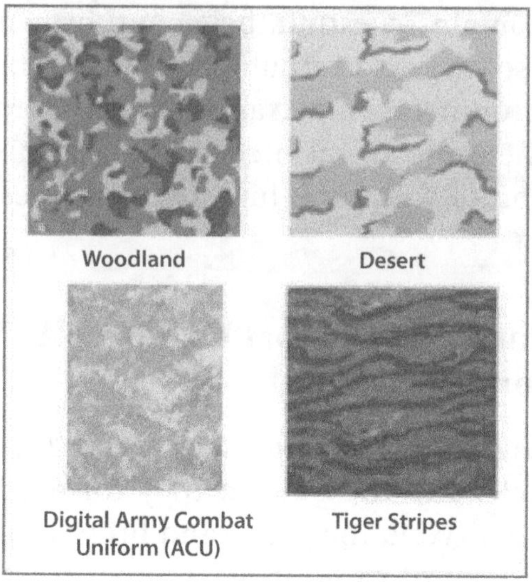

Figure 8.1 Various camouflage used in the military.

like the colors of tigers, hence, the name tiger stripe camouflage (Figure 8.1).

8.4.6 City or Urban Camouflage

This camouflage uses patterns with gray, black and white colors. These colors are those used in concrete, glass and steel buildings. This type of camouflage is also used for night operations. Figure 8.1 shows the principal military camouflage.

8.5 Active or Adaptive Camouflage

Active or adaptive camouflage refers to a group of technologies that allow an object to blend with its environment through the use of panels or coatings capable of altering its appearance, color, luminosity and reflective properties. With this, active camouflage provides the ability of perfect concealment from visual detection. One of the technologies that make this possible is that of organic light-emitting diodes (OLEDs). With the help of a camera, the object would not become invisible, but would mimic the environment that surrounds it.

In 2008, metamaterials appeared which have negative refraction indices for microwave frequencies. This means that the radiation that falls on them is not reflected or refracted, but curves around the object. Another possible technology designated as "phased array optics" allows a type of optical camouflage, when placing in front of an object the three-dimensional hologram of the background so that this object is hidden.

8.6 Nanomaterials for Advanced Camouflaging

Armies around the world are looking for a 21st century battle suit that must stop bullets, detect chemical and biological agents, monitor the vital signs of a wounded soldier, administer basic first aid and communicate with the headquarters. The military use of NT should lead to greater protection, more lethality, greater resistance and better self-support capabilities for future soldiers.

It is expected that with the aid of nanotechnology and high-strength nanocomposite plastics metal will be replaced and thus

- Reduce the weight.
- Reduce radar signature (RCS) of the military weapon platforms.
- Nano-sized silicon carbide particles for use in physical protection.
- Intelligent components, with built-in condition and load supervision sensors to improve the monitoring of important parameters. Biosensors can be used to monitor the health of a soldier.

- Nanotechnology enabled small sensors called nano-units, of which some simple types are available as "intelligent materials" that change in response to light or heat.
- Self-healing materials are also possible with nanotechnology.
- Shape memory alloys (that memorize form) are being used in the sleeves of combat aircraft.
- Active structures that adapt to changing conditions such as adaptive camouflage, suspension, flexible/rigid, nanofluids and shape memory alloys are examples of adaptive materials.
- In airplanes, suppression of the rotating wings vibration is especially a great design challenge. The nano-compound reduces vibrations.
- Radar absorption, camouflage and compound coatings reduce RCS.
- Nanomaterials can envelope the object, absorb radio waves emitted by the radar, and release them as thermal energy dispersed in space. By doing so, the material disguises the object, which makes its identification by radar difficult.
- Nanotechnology is also used to develop antiballistic structures, reactive armor of nanoparticles and shock-absorbing nanotubes.
- Surface coatings with nanoparticles make them harder, softer and/or more furtive.
- Nano-enabled paint is currently being developed as a special paint that makes unmanned aircraft, missiles or airplanes simply disappear or, to be more precise, they become very difficult to detect.
- The nanomaterials used in filters to remove selected impurities from fluids could become very low in cost and therefore ubiquitous, and result in many small but discrete concentrations of potentially

toxic impurities. "Argonide nanomaterials" filters can be used in war where water pollution is the biggest threat. A water purifier kit will be part of the survival kit in jungle operations to clean the water to drink [11].

- With the help of nanotechnology, electrochrome camouflage can be achieved, by which soldiers can disappear, and not be seen by the naked eye.
- Fabric made of electrochrome camouflage is capable of changing color instantly to blend with the environment.
- Nanofabrics made of nanopolymers are developed that are resistance to dust/deformation/fire.
- Recently, fabrics of nanopolymers are developed, which disintegrate bio/chemical agents of war.

8.6.1 Some Possibilities in the Near Future

"Nanobots" or small mobile robots have not yet been developed but are theoretically possible; and self-assembling nanomaterials that can be assembled into larger equipment. An artificial "electronic nose" which is trained to detect toxic gases and vapors for the detection of bio/chemical/nuclear weapons is being developed. Also in development are "nanowires" constructed from sub-micrometric layers of different metals, including gold, silver and nickel, which are able to act as "bar codes" to detect a variety of pathogens such as anthrax, smallpox, ricin and botulinum toxin. The approach could simultaneously identify multiple pathogens through their unique fluorescent characteristics. The capacity of soldiers is being increased with artificial muscles that could allow them to jump high walls, if they are injured or required to, by using nanotechnology and electricity. the artificial muscle flexes when shaken by electricity, and then relaxes when the electricity is turned off. So far, however, the reactions are too slow.

Nanoparticles are potentially important in ceramics and powder metallurgy and other similar applications. The strong tendency of small particles to form groups is a serious technological problem that prevents such applications. However, some dispersants such as ammonia citrate (aqueous) and oleic alcohol (nonaqueous) are promising additives for deagglomeration. They are nanoscale materials with structural characteristics of a dimension between 1–100 nanometers.

8.7 Summary

Multi-directional military applications of nanotechnology are being realized. One of the main aims of military research into nanotechnology is to provide the personnel with lightweight, strong and multifunctional materials for use in clothing, both for protection and to provide enhanced connectivity. Protection through the use of nanomaterials that can not only camouflage them but also provide improved medical and casualty care for soldiers, is getting the utmost attention of researchers.

References

1. Buzea, C., Pacheco, I., Robbie, K., Nanomaterials and nanoparticles: Sources and Toxicity. *Biointerphases*, 2, 4, MR17–MR71, 2007.
2. Cobb, M.D. and Macoubrie, J., Technology and applications: Public perceptions about nanotechnology: Risks, benefits and trust. *J. Nanoparticle Res.*, 6, 395–405, 2004.
3. Blechman, H. and Newmwn A., *DPM – Disruptive Pattern Material*, pp. 26–32, DPM Ltd, Frances Lincoln; London, UK, 2004.
4. Pendry, J.B., Schurig, D., Smith, D.R., *Science*, 312, 1780, 2006.
5. Schurig, D., Mock, J.J., Justice, B.J., Cummer, S.A., Pendry, J.B., Starr, A.F., Smith, D.R., *Science*, 314, 977, 2006.

6. Cai, W., Chettiar, U.K., Kildishev, A.V., Shalaev, V.M., *Nat. Photon.*, 1, 224, 2007.
7. Liu, R., Ji, C., Mock, J.J., Chin, J.Y., Cui, T.J., Smith, D.R., *Science*, 323, 366, 2009.
8. Gabrielli, L.H., Cardenas, J., Poitras, C.B., Lipson, M., *Nat. Photon.*, 3, 461, 2009.
9. Landy, N. and Smith, D.A., *Nat. Mater.*, 12, 25, 2013.
10. Chen, H.S., Zheng, B., Shen, L., Wang, H.P., Zhang, X.M., Zheludev, N., Zhang, B., *Nat. Commun.*, 4, 2652, 2013.
11. Tiwari, A., Military nanotechnology. *IJESAT*, 2, 4, 825–830, 2012.
12. Newark, T., Newark, Q., Borsarello, J.F., *Brassey's Book of Camouflage*, Brassey's, London, UK, 1996.

9

Applications of Nanotechnology in Aerospace

Madhuri Sharon

Walchand Centre for Research in Nanotechnology & Bionanotechnology, WCAS, W.H. Marg, Ashok-Chowk, Solapur, Maharashtra, India

In flying, the probability for survival is inversely proportional to the angle of arrival
Neil Armstrong

9.1 Introduction

The aerospace (Defense) industry is one of the most important heavy industries in the world. Countless companies rely on the ability to ship products and people around the world with a speed that can only be achieved by air. For acquiring the best aircraft performance, it is necessary to find an aircraft design that increases the efficiency (carrying payload). Therefore, some different methods have been proposed to increase the efficiency. These include nanomaterials. Nanostructure Science and Technology is a broad and interdisciplinary area of research with development activity that has been growing explosively worldwide in the past few years. It has the potential

Madhuri Sharon, et al. Nanotechnology in the Defense Industry: Advances, Innovation, and Practical Applications, (289–308) © 2019 Scrivener Publishing LLC

for revolutionizing the ways in which materials and products are created and the range and nature of functionalities that can be accessed. It is already having a significant commercial impact. Nanoscale generally refers to the size scale of 1 nanometer (nm) = 10^{-9} m. It refers to the matter whose length scale, in any dimension, is approximately 1 to 100 nanometers. The current trend is limited to use to some extent only in fighter aircraft, but this chapter depicts where and which type of nanomaterials can be used in almost the entire aircraft, including navigation system. Nanotechnology is a critical enabling technology for modern aviation and large-scale power generation with less visibility. Stealth technology is a subdiscipline of military countermeasures which covers a range of techniques used with aircraft in order to make them less visible (ideally invisible) to radar, infrared and other detection methods. Stealth technology (often referred to as "LO," for "low observability") is not a single technology but is a combination of technologies that attempt to greatly reduce the distances at which a vehicle can be detected. This aspect has been covered in Chapter 2 of this book.

The total spectrum of Defense Research and Technological activities required covers six areas:

- Applied Vehicle Technology
- Information Systems Technology
- Modeling and Simulation
- Studies, Analysis and Simulation
- Systems Concepts and Integration
- Sensors and Electronics Technology

For the defense arena, the applied vehicle technology encompasses or rather uses all other technically advanced technologies mentioned above, i.e., inclusion of information system, analysis as well as modeling and simulation, concept integration, sensors and electronic technologies.

9.2 Use of Nanomaterials in Different Areas of Aviation

For advancement of the aviation industry, the factors that are in consideration include lightweight aircraft, materials with multifunctional properties so as to have fewer or no repairs including self-healing, eco-friendly fuels, reduced fuel consumption, efficient communication systems, extended and safe life, faster, miniaturization, highly maneuverable, intelligence guided, smart materials with extraordinary mechanical and multifunctional properties.

Nanotechnology is being envisaged as a technology that is expected to have high impact in aerospace sectors as it can provide nanomaterials having high strength, low weight, multifunctional materials that will provide small and compact planes; and fully automated, self-guided, unmanned air vehicles for reconnaissance and surveillance. Nanoparticles have outstanding mechanical properties due to their perfect material structure; especially carbon nanotubes. Moreover, nanoparticles are not only strong but they offer a substantial reduction in weight, which is desirable for lightweight aircraft.

Nanomaterials can be primarily used in three areas of the aviation industry. These are structure of airframe, coating of airplanes and its engine parts, electrocommunication and detection of landmines. Nanomaterials used for these purposes are metals, ceramics, polymers, composite, etc., depending on the necessity of its use anywhere in the aircraft.

9.2.1 Airframe Structure

Since the onset of the development of manned flight in 1903 by the Wright brothers, the importance of lightweight material for an airplane's body and components of engine parts was considered. The Wright brothers had selected a heat-toughened alloy of aluminum. Since then it has been extensively used in aerospace engineering. With the aerospace advancements over the years, long-haul international flights came into existence, demanding engine components that are more durable and resistant to fatigue. To meet these requirements various alloys of aluminum have been developed. A very brief account of aluminum alloys that are in use nowadays are mentioned below and in Table 9.1.

9.2.1.1 Commonly Used Aluminum Alloys in Heavier Parts of the Aircraft

AA 2014 is a strong and hard alloy of aluminum, frequently used in the framework and internal structure of aircraft. Though it is appropriate for arc and resistant to welding, it has poor resistance to corrosion.

AA 2024 is the most common and popular high-grade alloy of aluminum used for the wings and fuselage of aircraft, because it has high tensile strength of approximately 470 MPa, high yield strength, and good fatigue resistance.

AA 5052 is a non-heat treatable alloy that has high resistance to corrosion, which is highly ductile and offers the highest strength. It is easily modeled into different shapes, including fittings and engine components.

AA 6061 is mostly used in homemade aircraft. It is extremely lightweight and relatively strong, which makes it perfect for the wings and fuselage. Moreover, it is easily welded and shaped.

AA 7050 is mostly used in military aircraft because it maintains strength in broad segments, has high corrosion resistance, so is more resistant to fractures. The fuselage and wing skins are made of this alloy.

AA 7068 is the strongest available alloy of aluminum that can withstand difficult conditions and attacks; hence, it is used in many military aircraft.

Table 9.1 Mechanical properties of some commonly used aerospace aluminum alloys.

Alloy	Temper	Density (g/cm)	Elastic modulus (GPa)	Yield strength (MPa)	Tensile strength (MPa)	Fracture toughness (MPa√m)
2014	T6	2.80	72.4	415	485	26.4
2024	T4	2.77	72.4	325	470	22.0
2219	T62	2.84	73.8	290	415	36.3
7050	T74	2.83	70.3	450	510	38.5
7075	T6	2.80	71.0	505	570	28.6

Source: *Aerospace Materials and Material Technologies. Volume 1: Aerospace Materials.* Indian Institute of Metals Series. N. Eswara Prasad, RJH. Wanhill. Springer Singapore.

AA 7075 is an alloy that has high levels of zinc, giving it excellent fatigue resistant, high strength properties like steel. It was a preferred alloy for fighter planes during World War II, when Mitsubishi A6M Zero fighters were used by the Japanese Imperial Navy on their carriers between 1940 and 1945. It is still used frequently in military aircraft.

9.2.1.2 Commonly Used Aluminum Alloys in Other Parts of the Aircraft

AA 2219 is an alloy which offers maximum strength at elevated temperatures. It was used for the external fuel tank of the first successfully launched space shuttle, Columbia. It has good weldability provided the welds are heat-treated to maintain resistance against corrosion.

AA 6063 is mainly used for architectural and aesthetic finishes and for finer details of aircraft intricacies.

AA 7475 is an alloy with high fatigue, fracture resistance and toughness. It is used in fuselage bulkheads of larger aircraft.

Al Lithium is a futuristic alloy, i.e., aluminum-lithium alloys have been made for the aerospace industry to decrease the aircraft weight and thus enhance the performance of the aircraft.

9.2.1.3 Aluminum Oxide Nanoparticles

The demand for aluminum is expected to be 80 million tons by 2025. More innovation in aluminum nanoparticle alloys is being researched by incorporating nanomaterials to be used for aircraft. Nanoparticles of aluminum oxide exhibit novel properties such as: (i) Thermal Conductivity: Nanoparticles of aluminum oxide are a hard spherical material with high stiffness and high thermal conductivity, which improves the thermal mechanical properties of the nano-epoxy composite. Conductive aluminum is an excellent conductor of heat and electricity. An aluminum conductor weighs approximately half as much as a copper conductor having the same conductivity. Aluminium-aluminum nanoparticles have a thermal and electric conductivity that is about 60% of copper, which accounts for its development as a conductor in the form of bars and tubes, which are used in numerous electrical applications. (ii) Corrosion resistance: Aluminium-aluminum nanoparticles react with the oxygen in the air to form an extremely thin layer of oxide, which provides excellent corrosion protection. The layer is self-repairing if damaged. Anodizing increases the thickness of the oxide layer and thus improves the strength of the natural corrosion protection. (iii) Recyclable: Aluminium-aluminum nanoparticles can be recycled indefinitely without losing any of their intrinsic qualities. (iv) Hard, wear-resistant, (v) Excellent dielectric properties from Dc to GH2 frequencies, (vi) High strength/stiffness, and (vii) Excellent size and shape capability.

Consequently, the aerospace industry is increasingly considering use of aluminum nanoparticles because the properties mentioned above can be enhanced by dispersing aluminium-aluminum nanoparticles into the fuel and the aircraft applications used. The fuel is used not only as a propellant but also as a coolant and for improving both the thermal conductivity and combustion enthalpy of the fuel is beneficial.

9.2.1.4 Nanomaterials for Airframe Structures

As discussed above, the properties of nanomaterials required for airframe structure should have light weight, high strength, high toughness, corrosion resistance, easy reparability and reusability, less maintenance and durability. Modern aviation design requires faster, miniature, highly maneuverable, self-healing, intelligence guided, smart, eco-friendly, lightweight and stealth systems with extraordinary mechanical and multifunctional properties. Materials that are being extensively researched and used are described below.

9.2.1.4.1 Carbon Nanotube (CNT)-Based Polymer Composites

CNTs in polymer composites have a wide range of Young's modulus, high specific strength, crash resistance and thermal performance. The polymers that are used in airframe structure are: CNT/Epoxy, CNT/Polyimide, and CNT/PP.

9.2.1.4.2 Nanoclay Reinforced Polymer Composites

These polymer composites provide barrier, thermal and flame retardant properties.

9.2.1.4.3 Metal Nanoparticles Incorporated Composites

These composites have electrostatic discharge and electromagnetic interference (EMI) shielding properties, which make them the futuristic material for making a structure which is resistant to lightning strikes.

9.2.2 Nanocoating

A coating is a covering that is applied to the surface of an object, usually referred to as the substrate. In many cases coatings are applied to improve surface properties of the substrate such as appearance, adhesion, wettability, corrosion resistance, wear resistance, and scratch resistance.

The application of nanotechnology in the protective coating has recently gained momentum as nanoscale materials have unique physical, chemical and physicochemical properties, which are expected to improve the corrosion protection in comparison to bulk materials.

Generally, components of coatings are: pigment, binder, additives, fillers and solvent. But nanocoatings are produced by using nanoscale particles of desired properties. Nanocoatings can be nanocrystalline, multilayer coatings with individual layers the thickness of nanometers and nanocomposites. SiO_2 nanoparticles are used as fillers. Coating can be (i) Functional Coating, which is self-cleaning, antifouling, easy to clean, soft feeling and bactericidal, or (ii) Self-Assembled Nanophase Coating, which is inspired by natural healing processes. Self-cleaning coatings have a special functional property, called the "lotus effect." That is, the ability of a biological surface to naturally repair itself. The self-cleaning property of lotus leaves are due to their specialized surface morphology that prevents dirt from forming an intimate contact with the surface, and the high hydrophilicity of their leaves that repels water. Consequently, as the water droplets roll onto the leaf surface, they carry away the contaminants.

Nanomaterials have been suggested as possible material for coating materials as they can be friction modifying agents, are carbides, nitrides, metals and ceramics, e.g., (i) SiC nanoparticles in SiC-particle-reinforced alumina, and (ii) Yttrium stabilized nano-zirconium.

To facilitate crack healing, high-temperature, high strength and creep resistance compared to monolithic ceramics, the following nanomaterials are being used: (i) TiN nanocrystallites embedded in amorphous Si_3N_4 are used for wear-resistant coatings; (ii) Nanocomposite coatings made of crystalline carbide, diamond-like carbide and metal di-chalcogenide, (iii) TiN for low friction and wear-resistant applications. (iv) Nano-graphite, nano-aluminium in polymer coating are used for electrostatic discharge, EMI shielding and low friction applications of aircraft surfaces.

Titanium dioxide has self-cleaning ability. It is hydrophilic due to its high surface energy; hence, water does not form drops on a surface coated with it, but a sealed water film instead.

When photocatalytic TiO_2 particles are illuminated with a UV light source (e.g., sunlight), electrons are promoted from the valence band (VB) to the conduction band (CB) of the particle. This creates a region of positive charge (h+) "holes" in the VB and a free electron in the CB. These charge carriers can either recombine or migrate to the surface, while the holes can react with the hydroxyl or adsorbed water molecules on the surface and produce different radicals such as hydroxyl radicals (OH·) and hydroperoxy radicals (HO_2·). For coating on a surface, TiO_2 nanoparticles are dispersed in an epoxy resin matrix.

Nanostructured coatings offer great potential for various applications due to their superior characteristics that are not typically found in conventional coatings. Self-assembled nanophase (SNAP) coatings are potential replacement for chromate-based surface treatments on aircraft aluminum alloys. The SNAP process is used to form thin, dense protective organic surface treatment coatings on Al aerospace alloys. The ability to design coating components from the molecular level upward offers tremendous potential for creating multifunctional coatings (Figure 9.1). SNAP coatings are used as part of a complete aircraft coating system to protect the aircraft's aluminum alloy from corrosion. The coating steps include, in order of application, surface preparation, surface treatment (SNAP), primer and topcoat.

The first coating for stealth application was developed in Germany during WWII on German submarines to absorb microwaves from radar. While no aircraft is totally invisible to radar, nanocoating aircraft makes it difficult for conventional radar to detect or track the aircraft effectively. Radar-absorbing

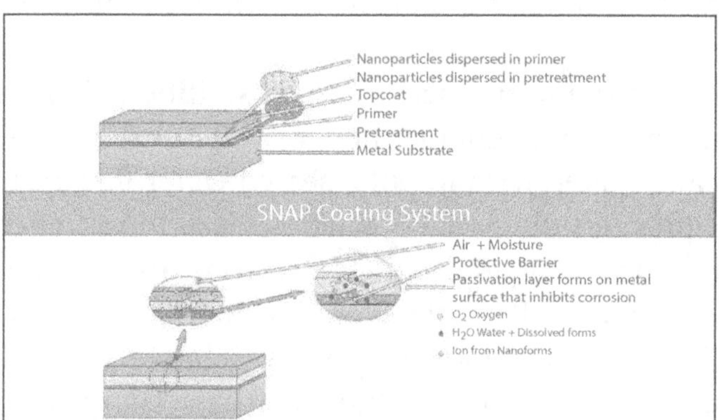

Figure 9.1 Schematic of Self-assembled nanophase (SNAP) system.

material (RAM) is basically paint coated on external surfaces. These paints are magnetic ferrite-based substance having dielectric ingredients such as carbon. RAM reduces the radar cross section, making the object appear smaller. In this the aircraft injects a stream of ionized gas which envelopes aircraft due to most of the radar waves being absorbed. Use of nanoparticles for coating aircrafts is for (i) heat-insulating the surface that protects aircraft engines from heat. Researchers at University West in Sweden have shown that this increases the service life of the coating by 300%. There are trials to spray-coat suspension plasma spray on the motor of aircraft with nanoparticles to shield the engine from heat. This will help in raising the temperature so that the efficiency is increased, fuel consumption is decreased and emission is reduced. There are trials to add plastic to the base material of the heat insulating layer that is ceramic. Addition of plastic is done to generate pores that make the material more elastic. Many efforts are on to make the layer elastic.

As claimed by FlightShield, a Swiss company that makes flight shield that is crystal clear nanocoating that chemically bonds to the aircraft paint surface and offers an ultra-smooth, high gloss and resistant protective layer coating. This coating stands between harmful contaminants and the paint, providing exceptional protection and gloss retention. FlightShield protects aircraft paint by providing a barrier to climate extremes, UV rays, deicing fluid, dirt and bugs. Moreover, FlightShield extends the life of aircraft paint by repelling harmful contaminants that would otherwise degrade it, giving paint very long life. The product cures and bonds to the painted surface, creating a long lasting and durable finish, as well as reduces cleaning expenses by repelling contaminants like dirt, oil, exhaust stains, deicing fluid and bugs. Simply put, the aircraft will stay cleaner for longer and therefore reduce cleaning costs.

The nanotechnology coating on the aircraft reduces carbon emissions by up to two per cent and also the fuel consumption.

Nanoshell is making aluminum nanoparticles coated engine parts for aircraft.

9.2.3 Aero Engine Parts

Nanoparticles are used for:

- **Super Capacitors** for airplanes are ceramic barium titanate and barium strontium titanate. They are NEMS (nanoelectromechanical systems) to develop a standard fuel management unit for fuel control in aero engines.
- **Gaskets and Sealants** used in aero engines are nano-graphite and nano-silica in rubber compounds.
- **Inhibiting Corrosion of Aluminum or Aero Structure** – Nano-chromium-based corrosion inhibitors are being developed.
- **Conductive Plastics** are used in various components of aircraft where electrostatic discharge of EMI shielding applications are required; they are nanopowder of copper, aluminum and iron.
- **Parachutes and Aircraft Arrester Textiles** are nanofibers and composites for nanofibers and polymers.

9.2.4 Aircraft Electrocommunication System

Use of nanoparticles in data storage media, transfer, and retrieval are at the moment in the extensive research stage and show great possibility in the very near future. With growth of IT markets and industries there is an increasing demand for next generation data storage technologies, which can move beyond Moore's Law, which is based on the article written by Gordon E. Moore [1].

According to him, the number of transistors on an integrated circuit will grow exponentially. His observations have withstood

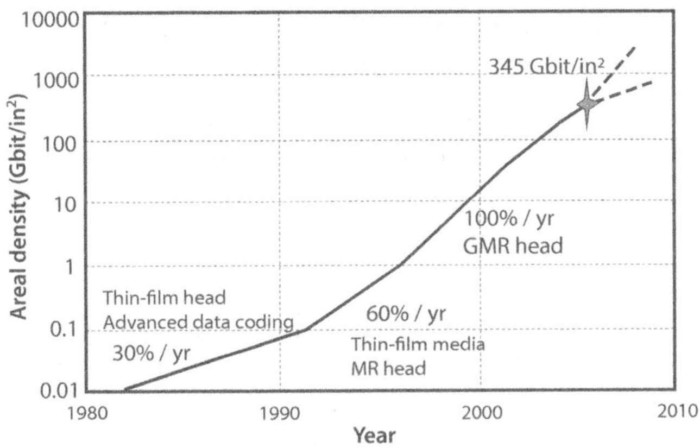

Figure 9.2 Data storage density for magnetic data recording. (Source: Xiaowei Teng, Meigan Aronson, 2011 [2].

the test of time since then. Figure 9.2 shows the trend in areal density of magnetic data storage during the past 30 years.

But after Richard Feynman's great proposition and Eric Drexler's practicality, a need for miniaturization of components and devices has taken center stage of research in the field of nanotechnology. Especially the quantum effect of the wave-like behavior of the electrons started to look into the designed functionality of the device. Some have predicted that the silicon technology is approaching its fundamental limits. Hence, miniaturizing the design for a data storage unit started with the demand to compress storage in smaller spaces, while keeping the energy demand low. To meet these demands nanotechnology has entered in the arena of miniaturized, faster and energy-saving communication system. Nanocomponents that are being tried for communication are:

- **Carbon Nanotubes (CNTs),** which have unique ballistic electron transport and a huge current carrying capacity, are being researched for their use in future nanoelectronics.

- **Fe_2O_3 & Fe_3O_4 Magnetic Nanoparticles** incorporated polymer films and composites can be used in patterning various data storage media. A patterned media is applied for ultrahigh density magnetic storage. This demands economical methods to prepare nanoscale patterned particles, fabrication of ultrafine, uniformly ordered nanoparticles, so as to obtain a theoretical data storage density as large as 1 T bit/in2 (1 Terabit = 1000 Gigabit) and a spacing of 25 nm between particles is required. If storage density is 10 T bit/in2, then the spacing required will be 8 nm; which is rather difficult to achieve by presently known nanofabrication methods. Therefore, low-cost, reliable patterning magnetic nanomaterials along with self-assembly and template-directed assembly of magnetic nanoparticles have been developed. This is an inexpensive and rapid process. These monodispersed magnetic nanoparticles can be organized onto the surface of substrates with two-dimensional (2D) or three-dimensional (3D) superlattice structures, patterned in such a way as to allow high frequency and highly parallel readout of the matrix of magnetic particles. Weak interaction forces, such as hydrogen bonding and hydrophobic interaction, steric repulsion, magnetostatic interaction, van der Waals interaction and Coulombic interaction, etc., support self-assembly. Here size, size distribution, shape of nanoparticles, and also the properties of solvents dominate the behavior of self-assembly. Using the same nanoparticles, different packing styles from hexagonal close packing, square packing, to linear chains have been achieved by selecting appropriate conditions.
- **Template-Assisted Assembly** - Nykypanchuk *et al.* [3], based on DNA self-assembly property,

used template-assisted assembly of the nanoparticles in a controlled manner with robust mechanical properties. A substrate functionalized with special molecules or pretreated to get particular surface structure is used as a template for the growth and assembly of various nanostructures.

- **Ceramic Nanoparticles** like barium titanate and barium strontium titanate are used for making supercapacitors.
- **Metal Oxide Particles** play a very important role in data storage and exchange bias. Hard disk uses superparamagnetic magnetic material for data recording. But there is a superparamagnetic limit, where ambient thermal energy reverses the recorded magnetization. To combat this defect, exchange bias magnetic materials are being developed. For this purpose, anti-ferromagnetically coupled, magnetic material with exchange bias properties is used. Desirable properties for an exchange bias material are: (i) A high Néel temperature (the temperature at which the thermal energy becomes large enough to destroy the macroscopic magnetic ordering within the material), (ii) A large magneto-crystalline anisotropy, and (iii) Good chemical and structural compatibility with ferromagnetic films (the materials record data directionally to represent either a 0 or a 1 binary digit). Antiferromagnetic oxides, like NiO and CoO, are significant exchange bias material.

There are efforts to use nano-sized CoO and NiO for their implication for the next generation data storage device. Metal oxide nanoparticles are synthesized using a "bottom-up" approach; such chemical methods performed in an aqueous or organic liquid phase. By this method it is easy to control the size and shape of metal oxide nanoparticles. Parameters that control it are reaction time, temperature and concentration of the reagents.

These metal nanoparticles show enhanced exchange bias effect [4]. To avoid the aggregation of nanoparticles caused by van der Waals force of attraction and/or magnetic interactions between magnetic nanoparticles existing in the nanoparticle system; a repulsive or stabilizing force is employed by adding surfactants, which provide a steric repulsion to increase the stability of nanoparticles [2]. Surfactants are organic compounds of long hydrocarbon chains.

9.2.5 Radar Technology for Detecting Landmines

One of the most challenging jobs in defense is to detect and safely remove buried landmines; because there are many possible types of landmines comprised of different materials and constructions. They could be buried in very different types of earth and be covered by various types of vegetation or obscurants. Hence, safe access is also an additional problem.

The most common landmine detection technology is use of metal detectors and detection using dogs. It is a slow and dangerous method as it can detect the explosive that is <1 m away. An innovative radar approach to this problem is proposed by Peichl *et al.* [5]. The problem one faces in developing a detection technology is that there are many different types of landmines composed of many different materials and constructions, which are buried at different depths, in different types of soil, and covered by different types of vegetation.

9.3 Possible Uses of Graphene in Aerospace

Graphene is a two-dimensional material that has the potential to improve aircraft performance, cost, and fuel efficiency. Graphene is potential nanofiller, when used as a polymer-based composite even at a very low loading it dramatically improves the properties such as tensile strength and elastic modulus, electrical and thermal conductivity, etc. Graphene/polymer composites are being tried for various potential applications due to their outstanding properties. Zhang *et al.* [6] prepared mechanically strong and electrically conductive graphene aerogels that possess the specific capacitance of 128 F g^{-1} with superior rate performance (at a constant current density of 50 mA g^{-1}) that can be used as electrode for electrochemical power sources; they are light weight (12–96 mg cm^{-3}), high conductivity (approaching 102S m^{-1}), and have large BET surface area (512 m^2 g^{-1}), ample volume (2.48 cm^3 g^{-1}) and hierarchically porous structure. The aerogels were prepared by two methods: (i) supercritical CO_2 drying and (ii) freeze drying of graphene hydrogel precursors obtained from heating the aqueous mixture of graphene oxide with L-ascorbic acid without stirring. Moreover, due to their mechanical property, the graphene aerogels can support more than 14000 times their own weight, nearly twice the amount supported by CNT. Because of thier light weight and large surface area, along with their electrical and mechanical properties, Siochi [7] from NASA, in her article "Graphene in the Sky and Beyond," has suggested their use in many aerospace applications.

The excellent thermal properties of graphene have compelled a collaboration between the Graphene Flagship and the European Space Agency for testing graphene for two different space-related applications, which has shown very promising results in improving the performance of loop heat pipes, thermal management systems used in aerospace and satellite applications. The main element of the loop heat pipe is the metallic wick, where heat is.

Ray Gibbs, CEO of Haydale Technologies Inc., has mentioned that using graphene they have developed electrically conductive epoxy resins which can significantly increase the electrical conductivity of the carbon composite materials used in the construction of fuselages to such a level that it would hopefully negate the use of copper mesh (which is employed for the prevention of damage caused from lightning strikes). It's essential to have a good conductor. Copper mesh is currently used in the body of the aircraft. Many composite materials used in aircraft, such as trays, overhead storage lockers, toilet fixtures, etc., could all be made from graphene composite materials and reducing weight, which is the constant goal of the aircraft industry, without compromising safety.

Recently, in March 2018 a new paper on the potential of graphene within the aerospace industry has been launched by the Aerospace Technology Institute (ATI) and the National Graphene Institute (NGI) at The University of Manchester in the U.K. According to the paper, "aircraft safety and performance could be significantly enhanced by including atom-thin graphene in the materials used to manufacture aircraft." Use of graphene is also expected to reduce the weight of the material, contributing to improved aircraft fuel efficiency.

9.4 Stealth Technology

The role of nanotechnology in stealth is discussed in detail in Chapter 2 of this book.

9.5 Summary

This chapter has presented the potential of nanomaterials in the Aviation Sector, including stealth technology. The use of nanotechnology offers light weight, high strength, high toughness, corrosion resistance, easy reparability and reusability, less maintenance, durability, and cost-effectiveness as it carries increased payloads and is safer for protecting the structures and surfaces of the aircraft from harsh environments than conventional technology.

References

1. Moore, G.E., Cramming more components onto integrated circuits. *Electronics*, 38, 8, 117, 1965.
2. Teng, X.W. and Yang, H., Effects of surfactants and synthetic conditions on the sizes and self-assembly of monodispersed iron oxide nanoparticles. *Journal of Materials Chemistry*, 14, 774–779, 2004.
3. Nykypanchuk, D., Maye, M.M., van der Lelie, D., Gang, O., DNA-guided crystallization of colloidal nanoparticles. *Nature*, 451, 7178, 549–552, 2008, doi: 10.1038/nature06560.
4. Inderhees, S.E., Borchers, J.A., Green, K.S., Kim, M.S., Sun, K., Strycker, G., Aronson, M.C., Manipulating the magnetic structure of Co Core/CoO shell nanoparticles: Implications for controlling the exchange bias. *Phys. Rev. Lett.*, 101, 11, 117202, 2008.
5. Peichl, M., Schreiber, E., Heinzel, A., Dill, S., Novel imaging radar technology for detection of landmines and other unexploded ordnance. *Eur. J. Secur. Res.*, 2, 23–37, 2017, doi: DOI 10.1007/s41125-016-0011-3.
6. Zhang, X., Sui, Z., Xu, B., Yue, S., Luo, Y., Zhan, W., Liu, B., Mechanically strong and highly conductive graphene aerogel and its use as electrodes for electrochemical power sources. *J. Mater. Chem.*, 21, 6494–6497, 2011. www.rsc.org/materials.
7. Siochi, E.J., Graphene in the sky and beyond. *Nat. Nanotechnol.*, 9, 745–747, 2014.

Index

Also of Interest

Check out the other titles in the "Advances in Nanotechnology & Applications" series

Nanotechnology in the Defense Industry: Advances, Innovation, and Practical Applications
By Madhuri Sharon, Angelica Silvestre, Lopez Rodriguez, Chetna Sharon and Pio Sifuentes Gallardo
Published 2019. ISBN 978-1-119-46012-1

The History of Nanotechnology: From Prehistoric to Modern Times
Edited by Madhuri Sharon
Published 2019. ISBN 978-1-119-46008-4

Fundamentals of Electrocatalyst Materials and Interfacial Characterization
Energy Producing Devices and Environmental Protection
By Nicolas Alonso-Vante
Published 2019. ISBN 978-1-119-46007-7

Behaviors and Persistence of Nanomaterials in Biomedical Applications
By Domenico Cassano and Valerio Voliani
Published 2018. ISBN 978-1-119-41827-6

Carbon Dots as Theranostic Agents
By Madhuri Sharon and Ashmi Mewada
Published 2018. ISBN 978-1-119-46010-7

www.ingramcontent.com/pod-product-compliance
Lightning Source LLC
LaVergne TN
LVHW020420301125
826312LV00002B/5
9781119460121